exploring
COMMUNICATIONS

by

Richard D. Seymour
Assistant Professor
Department of Industry and Technology
Ball State University
Muncie, Indiana

John M. Ritz
Associate Professor
Department of Vocational and Technical Education
Old Dominion University
Norfolk, Virginia

Florence A. Cloghessy
Editor—Technology
Goodheart-Willcox Company
South Holland, Illinois

South Holland, Illinois
THE GOODHEART-WILLCOX COMPANY, INC.
Publisher

Library of Congress Catalog Card Number 86-27009
International Standard Book Number 0-87006-539-4

123456789-87-0987

Library of Congress Cataloging in Publication Data

Seymour, Richard D.
  Exploring communications.

  Includes index.
  1. Communications.
  I. Ritz, John M. II. Cloghessy, Florence A.
  III. Title.
  NA7115.K46 1984   728.3'028   86-27009
  ISBN 0-87006-539-4

INTRODUCTION

EXPLORING COMMUNICATIONS will introduce you to the communication process. You will explore this process as it is used to exchange ideas and feelings. You will learn that the designing, coding, transmitting, and receiving steps are used in every message.

EXPLORING COMMUNICATIONS also examines the technical devices that aid human communication. These devices range from simple to complex. You will learn how each device makes communication possible, for humans and machines. The telephone, printing press, satellite, and computer are some devices studied in the text.

The impact of communication on societies and cultures of the world is also discussed in EXPLORING COMMUNICATIONS. You will discover the purposes, influences, and uses of communication in our society. You will also gain knowledge about the variety of careers made possible by the communications industry.

EXPLORING COMMUNICATIONS focuses on three areas of the communications industry. These areas are technical graphics, printed graphics, and electronic communications. Each area has a chapter of introduction. In addition, there are chapters dealing with technical devices used in each of these areas.

EXPLORING COMMUNICATIONS will help you in gaining a broad background in the study of communications. You will be able to use knowledge gained from this study in many other areas of your learning.

Richard D. Seymour
John M. Ritz
Florence A. Cloghessy

3

CONTENTS

1 | Introduction to Communications

The information given in this chapter will enable you to:
○ *Define communication technology.*
○ *Describe how communication devices have extended our senses.*
○ *Explain how people and machines transfer information.*

Forms of communication have been used since our world was first inhabited. Early civilizations expressed themselves using signs and symbols. Later language systems came about so we could talk to each other. Language systems developed differently in various parts of the world. Today we have many languages. Some of these include English, Spanish, Chinese, and French, Fig. 1-1. The use of languages is just one example of humans exchanging information. We call this

COMMUNICATION
ENGLISH

MITTEILEN
GERMAN

交通
CHINESE

EN COMMUNICATION AVEC
FRENCH

COMUNICAR
SPANISH

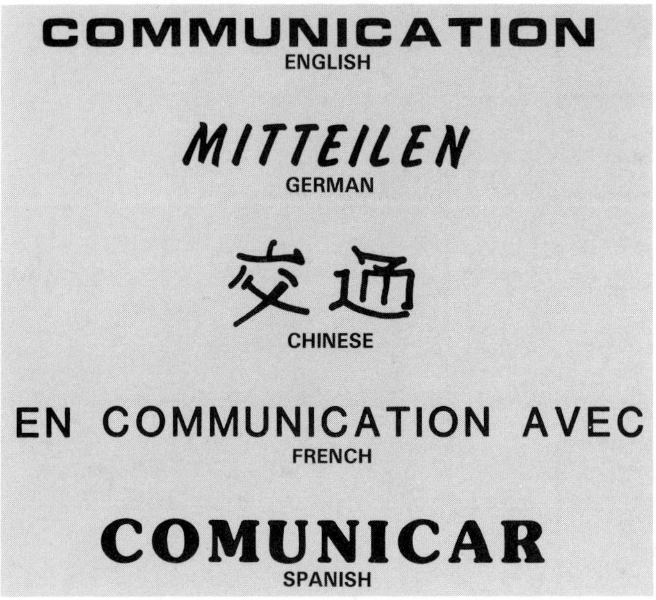

Fig. 1-1. People throughout the world communicate in their native languages.

communication.

Humans are not the only inhabitants of the world that exchange messages. Studies show that animals and plants communicate also. For example, deer often raise their tails to indicate danger. Dolphins are famous for their ability to communicate with each other. Many forms of plant life react to natural signals too.

However, we are interested in the technical side of communication. We will explore how people transmit important ideas and feelings. This book will focus on how information is changed for human use.

Various technical systems exist to aid humans in communicating. Some devices help create books and newspapers. Others let us enjoy radio and TV programs. Computers often assist with processing important information. Many forms of communication technology improve our daily lives.

COMMUNICATION TECHNOLOGY

Communication is the process of exchanging information. The word communication comes from the Latin word "com-muno." This means to transmit, pass along, or make known. Talking and listening are easy ways to pass along information. We use these methods every day to communicate. Sight is another simple form of exchanging messages. What we see gives meaning to what we know and hear. The saying "a picture is worth a thousand words" is very famous. We use our sight to read books and watch television. See Fig. 1-2.

As we mentioned earlier, many technical devices aid us in communicating. The telephone is a common example. People often talk to others around the world. Television is another important communication device. Traffic lights, door bells, and clocks are others. How would we communicate without these items?

Fig. 1-2. This family is enjoying a TV program on their television set. What TV programs do you like to watch? (RCA)

Communication technology includes the study of technical processes used to exchange information. Several processes make up this procedure. When people talk about communication technology, they are often only referring to one procedure. For example, some people refer to the sending and receiving of messages. Many describe the use of symbols, signs, and words. Others explain how messages are transferred and stored for future use. Most people feel that feedback of transmitted signals is important to most systems. All of this can become confusing.

To fully understand the meaning of communica-

tion technology a definition is needed. This definition must fully explain the process that takes place. It should include the act of communicating between humans and machines. Therefore, we have developed a general yet complete definition. COMMUNICATION TECHNOLOGY is the process of transmitting information from a source to a destination using codes and storage systems. This is illustrated in Fig. 1-3.

USING OUR SENSES

Most of us have five senses. These include hearing, touching, seeing, smelling, and tasting, Fig. 1-4. Hearing and seeing are the two senses most used by humans. Hearing is the sense that lets us receive messages by sound waves. Seeing is the human sense associated with our eyes. We receive visual information in this manner. Touch is a third important sense. We feel pressure on the body by human contact. With this sense we learn about temperatures, shapes, and textures. Smelling is the ability to recognize various scents and odors. The fifth sense is taste. We enjoy many foods and beverages because of their pleasing taste.

We often use all five senses at once. If someone is cooking nearby, we can describe the event with all our senses. Many times, however, only one sensory system is used. Talking on the telephone is an example. Reading a book (without pictures) is another. More often, a combination of senses

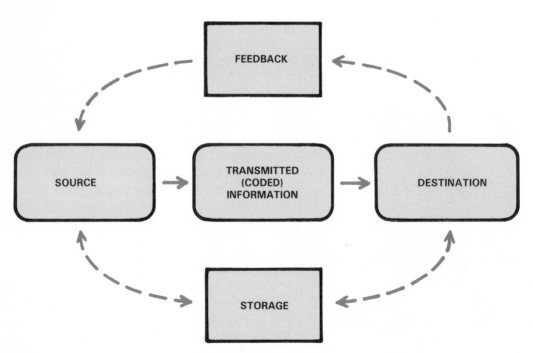

Fig. 1-3. Technology has helped humans communicate more efficiently. Our systems of exchanging information includes these elements.

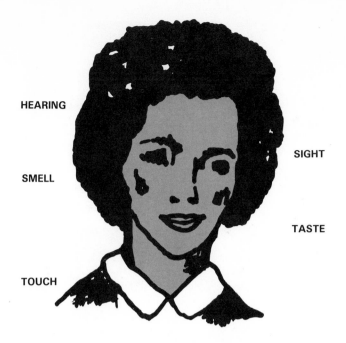

Fig. 1-4. We use our five human senses when communicating with others.

HEARING

SMELL

TOUCH

SIGHT

TASTE

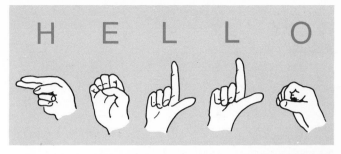

Fig. 1-6. Individuals who are deaf communicate with hand signals called "signing."

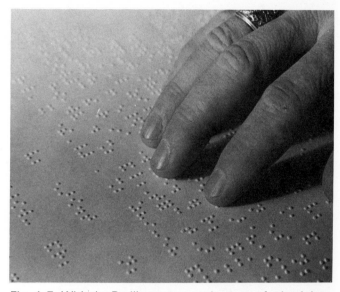

Fig. 1-7. With the Braille system each group of raised dots represents a letter or number.

are used when communicating. When we watch and listen to a movie or TV program, we use a combination of senses, Fig. 1-5.

These simple sensory exchanges of information are not available to all people. Handicapped people need other methods for communication. For those not able to hear or speak, there is a language called signing, Fig. 1-6. SIGNING uses the hands and fingers to spell out words. Another system has been developed for the blind. It is known as BRAILLE. All letters of Braille words are printed as raised dots. By moving the fingers across a page, Fig. 1-7, a blind person can "feel"

Fig. 1-5. Most of us enjoy watching movies on TV or at the theater. When did you last watch a feature film with your friends or family? (Zenith Electronics Corp.)

the words. This makes up for the lack of sight.

People must use their senses to communicate. The five senses help us pick up and transfer messages. Many items are designed to aid these sensory abilities. Examples include radios and magnifying glasses. The reason for communication technology is to allow our senses to receive small or faint messages. We will see how this technology actually "extends" our senses.

EXTENDING HUMAN SENSES

We hear others speak to us with our sense of hearing. With radio or telephones, we can hear them over long distances. These two devices extend the human ability of hearing. This is an example of how technology assists us.

Inventors have developed many instruments to aid in communication, Fig. 1-8. Look around any school building. Teachers use a PA (public address) system for general announcements. Cheerleaders have megaphones for sporting events, Fig. 1-9. Members of the school band play

Fig. 1-8. Large signs and alarm systems aid humans in communicating with others. Can you name other communication instruments found nearby?

process data is an excellent example. A company will record vital information on computer disks, Fig. 1-10. The computer might be connected to other machines across the country. Information

Fig. 1-9. This cheerleader yells to the crowd with her megaphone.

Fig. 1-10. Modern computers store information as electrical signals on magnetic tape or flat disks. This disk drive unit is connected to the monitors shown in the background. Any information needed is displayed on the screen and then stored for later use. (Sperry Corporation)

musical instruments. Bulletin boards feature announcements of school events.

These illustrations of communication are among the simplest found. Other forms are more complex. They involve entire systems of recording and transmitting messages. Using a computer to is relayed by wires or radio signals. This advanced form of communication involves an entire system. We are now ready to further explore communication systems.

INFORMATION TRANSFER

There are four major types of communication systems. One is human-to-human communication. Others include human-to-machine, machine-to-human, and machine-to-machine. A fifth system of communication is known as supplemental. These methods of communication are different than human and machine systems. Supplemental forms of communication will be discussed later. The five basic communication systems are shown in Fig. 1-11.

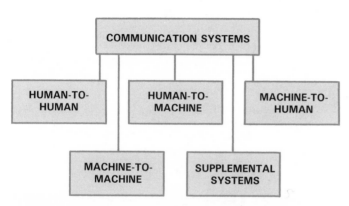

Fig. 1-11. Information is transferred between machines and humans in several ways. Can you think of examples of each system?

HUMAN-TO-HUMAN SYSTEMS

When we speak or write personal notes, we communicate with other people. This is the simplest method of exchanging information. It is called human-to-human communication. This type of exchange uses only basic forms of technology. Pens, pencils, paper, and signs are simple aids. Languages and signals are others. In human-to-human communication, no machines assist the sensory system in receiving messages. Therefore, setting a combination lock is not human-to-human communication. Neither is programming a computer or using an alarm clock. However, writing on a chalkboard does fit this description, Fig. 1-12.

HUMAN-TO-MACHINE SYSTEMS

Machines help humans produce many goods and services. But people have to operate these devices, Fig. 1-13. To complete a school assignment, we

Fig. 1-12. Teachers often use a chalkboard to provide a visual message for students. Graphic images help us better understand facts and details.

Fig. 1-13. These dispatchers use television monitors, telephones, and computer printouts to manage the flow of railroad trains around the country. (Santa Fe Railway)

often use a typewriter. In doing so, we communicate to the machine with our fingers. The final product is the typewritten page. Communication between humans and machines is quite common. We dial telephones, turn on lights, and operate power tools. The list is endless in our technological society. Human-to-machine communication extends our limited capabilities.

MACHINE-TO-HUMAN SYSTEMS

A third system of transferring information is from machines to people. This occurs when machines tell us how they are functioning, Fig. 1-14. We also retrieve messages from machines. Tape players and stereos are common examples. A voice or song can be recorded on a tape or disk. A machine can then reproduce the recorded signals so we may hear and enjoy it.

In industry a machine will often make unusual noises when it needs repairs. This noise then alerts the person in charge that repairs are needed. Other machines that communicate with humans include banking machines and warning buzzers. Around the home we find smoke detectors and alarm clocks. The gasoline gauge in a car warns us when the tank is nearly empty, Fig. 1-15.

The primary reason machines communicate with people is to obtain information. These messages often save time and energy. At the same time, other machines entertain us. Machines have assisted humans in many areas of communication.

MACHINE-TO-MACHINE SYSTEMS

Machines communicating with other machines did not exist on a large scale until recent years. Computers are an example of this type of communication. They often direct the operation of other machines, Fig. 1-16. A term we relate to this process is CYBERNETICS. This means one machine running another.

Many forms of machine-to-machine communication are found in the home. Temperature controls (thermostats) on furnaces maintain warmth in rooms. They also control air-conditioners during summer heat. These devices turn off the air-conditioner or furnace when the temperature is too hot or too cold. Timers and switches on your stove work the same way.

Communication between machines is important. It helps us get things done when we need them completed. Some controls turn machines on and off. Others change speeds. Many devices stop machinery in times of danger. These systems pre-

Fig. 1-14. The operation of this machine is monitored by a computer. The work is then displayed on a screen for workers to observe. (AccuRay Corp.)

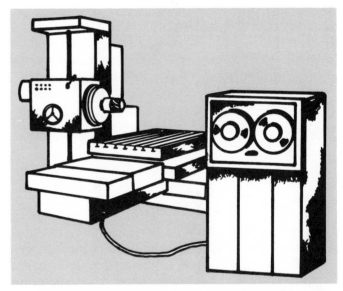

Fig. 1-16. Many factory machines are directed by other machines. A common example is when a computer guides the operation of an NC (numerical control) machine.

Fig. 1-15. The dashboard of a typical automobile includes many communication instruments. How many can you identify in this photo? (Chrysler Corp.)

SUPPLEMENTAL SYSTEMS
Extrasensory Perception (ESP)
Extraterrestrial
Animal
Plant
Mineral

Fig. 1-17. Types of supplemental systems.

vent injury to people. They also turn machines off when not needed. This saves energy.

SUPPLEMENTAL SYSTEMS

Common forms of communication exist between humans and machines. We can also identify other systems for exchanging information. These methods of communicating do not fit into our four general categories. Therefore, they rate their own grouping. Examples of supplemental systems include extrasensory perception (ESP), extraterrestrial, animal, plant, and mineral systems, Fig. 1-17.

Extrasensory perception is thought to exist outside normal human senses. People thought to have ESP claim to have mind-reading abilities. They attempt to predict events before they happen. Some scientists believe in this form of communication. Others doubt if it can be proved. This method of communication is uncertain at best.

Information may also be exchanged between planets or spacecraft. Messages received from outside the earth are called EXTRATERRESTRIAL. This sytem is often shown in science fiction movies. Many scientists are searching for extraterrestrial messages. They use large radio devices to scan the skies, Fig. 1-18. Scientists hope to one day communicate with life on other planets.

Plants and animals communicate also. Research has revealed many interesting discoveries. For example, some animals can communicate by using common signals. Lightning bugs exchange information with their light signals. Honeybees fly in special formations to communicate. Even plant life reacts to natural signals. Some researchers believe plants grow better when exposed to music.

A final communication system involves different types of minerals. Rocks and other earth substances communicate to humans when aided by instruments. Magnets attract iron ore and other metals. Radioactive materials send off dangerous waves. A compass (pointing towards the north pole) is an example of mineral communication. Scientists often use complex devices to analyze minerals. This helps us learn about their qualities.

SUMMARY

This chapter has introduced the meaning of communication and communication technology. We rely on our senses to communicate with others. Technology has extended the capabilities of our senses. We have learned what must occur in the communication process. Information (in the form of messages) is transmitted. These messages can also be stored for later usage. Finally, we reviewed the major communication systems.

KEY WORDS

All the following words have been used in this chapter. Do you know their meanings?

Braille, Communication, Communication process, Communication technology, Cybernetics, Extraterrestrial, Human-to-human systems, Human-to-machine systems, Information, Machine-to-human systems, Machine-to-machine systems, and Supplemental systems.

TEST YOUR KNOWLEDGE — Chapter 1

(Please do not write in the text. Place your answers on a separate sheet.)

MATCHING QUESTIONS: Match the definition in the left-hand column with the correct term in the right-hand column.

1. __ A reading system for the blind.
2. __ The process of one machine running another machine.
3. __ A language that uses the hands and fingers to spell out words.
4. __ Messages received from outside the earth.

a. Braille.
b. Extraterrestrial.
c. Cybernetics.
d. Signing.

Fig. 1-18. Large radio telescopes continually scan the sky for signals from other planets.

5. Define communication technology.
6. The five senses include _____, _____, _____, _____, and _____.
7. Identify the five systems of information transfer. Give an example of each system.

ACTIVITIES

1. Collect five definitions of communication from other books.
2. Select three technical methods of communication (for example, a home computer, a telephone, and a newspaper). Illustrate them in diagram form. This diagram should be based upon the communication model found in Fig. 1-3.
3. Invite an individual from a telephone company, local TV or radio station, or other communication business to speak to your class about the work done by the company.
4. Invite a teacher who works with deaf and blind individuals to visit your class. Ask the instructor to demonstrate the signing and Braille systems of communication.
5. Make a collage illustrating the communication process, using either photographs or clippings from magazines.
6. Draw the symbols used for Olympic events or for international road signs. Make these into mobiles or bulletin boards.
7. Draw maps showing where students live.
8. Take a quick trip around the classroom or school. List all the communication devices you find and the type of information transfer system used.

2 | History Of Communication Technology

The information given in this chapter will enable you to:
O *Briefly describe the three major developments in early communication technology.*
O *Identify how science has assisted in the development of communication devices.*
O *List the most significant communication devices invented.*

Over the years, many technical developments have helped extend our senses. In particular, they aid our senses of sight, sound, and touch. The progress in this area has been great. Thousands of years ago humans communicated through basic languages and drawings. Now we are able to make light, sound, and feelings more intense. This is

Fig. 2-1. Telephones let us communicate over long distances. Invented in 1876, they remain a popular communication device. (Graphic Arts Technical Foundation)

called AMPLIFICATION. It lets us send signals faster and farther than ever possible. It is done with the aid of electronic devices, Fig. 2-1.

Most basic communication devices have been modified only slightly. This chapter will focus on the most common developments of communication technology.

EARLY COMMUNICATION DEVELOPMENTS

Society uses communication devices for a variety of purposes. The history of this technology is exciting. You will be surprised to learn how many early developments are still used today. We will see how humans have improved their ability to communicate.

SPEECH

The human species began communicating with their voices 25,000 years ago. At first only grunts and gestures were used. Then these simple actions developed into a system of spoken words. It is uncertain how so many languages evolved. One language probably began with each new society. Today speech is the basis for most communication systems. Telephone, radio, and personal communication all depend on verbal communication (talking).

DRAWING

Early humans also learned to communicate through drawings. Our first signs of graphic communication are cave drawings. Drawings by cave dwellers developed into what we call HIERO-GLYPHICS, Fig. 2-2. This form of drawing used pictures to communicate ideas. Cave drawings

Fig. 2-2. Early civilizations developed their own language systems. One society used symbols like these to communicate by graphic means.

were often pictures of animals or tools. Simple maps were often also drawn. Early drawings were further improved with the use of instruments. Today this method of communication is known as DRAFTING.

Communicating through graphics is very useful today. For example, new developments in computer graphics have meant new standards of excellence, Fig. 2-3.

Fig. 2-3. A computer is used to prepare an engineering drawing. We call this computer graphics. (General Motors Corp.)

PRINTING WITH MOVABLE TYPE

Before the 1400s, the reproduction of drawings and writings was very slow. This work had to be done by hand. Therefore, printed materials (like books) were very expensive. Only the wealthy could afford printed materials. But in 1450, Johann Gutenberg invented movable type for printing, Fig. 2-4. The press worked in the following manner. The paper to be printed was placed over the inked type. A smooth, flat plate, called a PLATEN, was pressed down on the paper. The platen was lifted, and the printed paper removed. Gutenberg's press helped change the course of history. Humans could now communicate more easily. The ability to print words and pictures caught on quickly. Many new ideas in printing technology have evolved since Gutenberg's day.

HOW SCIENCE AIDS COMMUNICATION

Gutenberg's invention of movable type came at an important time. A social revolution was beginning in Europe. It was called the Renaissance. Between the 15th and 17th centuries, people gained new freedoms. They wanted to expand their knowledge. Beliefs from medieval times suddenly changed. During this period, sailors left Europe to discover new lands. People also began to study science and mathematics. This new research

Fig. 2-4. Gutenberg invented movable type in the 1400s. Letters such as these could easily be assembled into words.

directly affected communication technology. New instruments were developed for research use. These devices helped measure and record experiments. The thermometer, microscope, and pendulum clock are common examples, Fig. 2-5.

By the 18th century, scientists began important experiments in chemistry. Several French scientists used their knowledge of chemicals to produce the first practical photograph. Reports of this work (printed on a Gutenberg press) circulated Europe. New ideas spread among researchers quickly. The development of improved systems grew quickly.

Other scientists were experimenting with electrical principles. In 1831, Michael Faraday produced a flow of electrical current with a simple magnet and coil. Further research led to a basic understanding of electricity. This led directly to the invention of many now common items, such as light bulbs and compact hairdryers.

Industry grew quickly in the western world between 1750 and 1900. Factories were built across the land. Steam and electrical power came into common use. Research efforts increased in many areas. Scientific knowledge combined with technical growth to improve many devices. These items helped with the transfer of new information. Messages could now be spread over greater distances at faster speeds. This resulted in greater efficiency.

A number of early inventions are still in use today. Others have been replaced by more useful items, Fig. 2-6. We will now look at these important communication developments.

ELECTRIC TELEGRAPH

The electric telegraph was invented in the middle 1800s. This instrument came from the practice of signaling others with flags or semaphore. The telegraph uses electrical current in a coil to attract an iron lever, Fig. 2-7. This creates the

Fig. 2-5. Have you ever used a microscope? With the aid of this instrument, it is possible to see very small details. (Bausch & Lomb)

Fig. 2-6. The telephone, radio, and television are common devices used in communication.

Fig. 2-7. The telegraph was once an important communication device. Transmission of telegrams (messages) is still popular today.

familiar "clicking" sound.

With the telegraph, electrical pulses could be transmitted over long distances. But a system of understandable signals was required to make the telegraph more useful. The solution to this problem was introduced in 1837. Samuel Morse devised a code using long and short sounds, or "clicks," Fig. 2-8. With Morse's code, messages were sent along wires by spelling out each word. This system was still slower than regular talking. However, it made long-distance exchange of information more efficient.

TELEPHONE

The telegraph was a large step in improving communication technology. But it could not transmit a human voice. Many people experimented with electricity to find a way around this problem. A solution finally appeared in the late 1800s. Alexander Graham Bell invented the first working telephone in 1876. Using a diaphragm excited (moved) by an electric current, voice tones could be transmitted. The age of electronic communication had arrived! Today's telephone systems extend our voices worldwide. Automatic switchboards, microwave transmissions, satellites, and optic fibers send telephone signals around the globe, Fig. 2-9. This communication system continues to play a major role in modern society.

RECORDER

The printing press allowed visual information to be stored. Both words and symbols (pictures) could be recorded for later use. But the telegraph and telephone were different. They functioned by exchanging electrical impulses. There was no way to record and store these signals.

People often want to preserve voices and musical tunes. Therefore, an invention to record these messages was sought. In 1877, Thomas Edison developed the PHONOGRAPH. It could

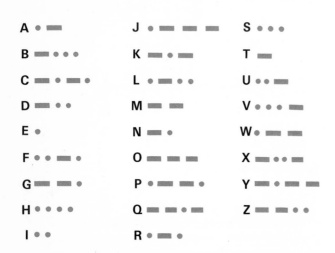

Fig. 2-8. Morse Code is a series of long and short sounds transmitted by telegraph devices. This "language" represents a common type of coding system.

Fig. 2-9. Radio waves are transmitted across the country by a complex microwave network.
(Rockwell International Corp.)

record clear signals, and play them back at a later time, Fig. 2-10. Other forms of recorders soon followed. Among the first was the tape recorder. People could easily "save" messages on tape at home or work. Cassette recorders have since become very common. In various industries, recorders are in wide use. Radio and television programs are usually recorded for later broadcast. Many firms produce commercial albums and tapes. These can be enjoyed at home on stereo equipment. The development of a recorder was significant in communication technology.

Fig. 2-10. Thomas A. Edison invented the first practical phonograph. Today's models have many convenient features compared to earlier models.

PHOTOGRAPHY

The growth of sound communication was paralled by the growth in photography. The invention of photography occurred in the early 1800s. The first practical photograph was called a DAGUERREOTYPE. This name came from the inventor of the process, Louis Daguerre of France. Producing daguerreotypes was costly and very complicated. However, the process remained popular until the 1870s.

Many improvements since then have made modern photography possible. Camera design, film, lenses, and color chemistry are among the most important improvements. Today people can take pictures for a small cost. Color slides and prints are used by both business people and hobbyists. Modifications in photography have led to motion pictures, VCRs, and other reproduction processes.

RADIO

In 1895, Guglielmo Marconi (of Italy) devised the first "wireless" telegraph. His work combined the efforts of many other people. Along with several original ideas, Marconi paved the way for radio broadcasting. Messages could finally be transmitted over long distances without wires. This was accomplished by using radio waves to carry signals. These signals were sent through the atmosphere to radio receivers, Fig. 2-11.

The most attractive characteristic of radio communication was its lack of wires. Radio waves could be sent to distant points easily. More importantly, news and other programs could be transmitted over large areas. Information could reach people around the earth.

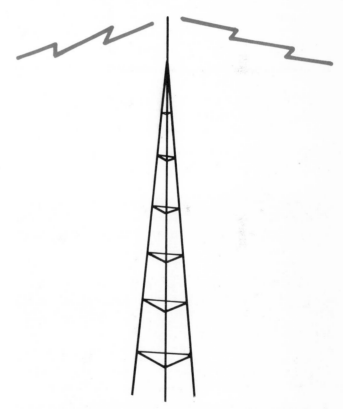

Fig. 2-11. Radio waves (signals) travel through the air to receiving units (radios). This is an effective means of communication; about 99% of all households have at least one radio.

MOTION PICTURES

Many inventions led to the development of projected, moving images. The first person to produce (and project) a motion picture is not known. However, much of the credit is given to Thomas Edison. His work with a kinetoscope was a key development in this area.

The KINETOSCOPE was a machine used for viewing "moving" pictures. A group of photographs were put on a single roll of film. This roll was moved in a constant motion past a light source. When watched through a viewing hole, it looked as if it were moving, Fig. 2-12.

As time passed, major improvements in motion pictures occurred. Projectors became more powerful, Fig. 2-13. Sound and color added reality.

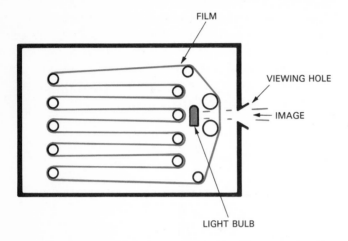

Fig. 2-12. The first "motion" pictures were made possible with the kinetoscope. As single photographs passed the viewing window, action was created. Viewers had to look into the machine to watch the movie show.

Fig. 2-13. Most schools have movie projectors like this for instructional films. (Bell & Howell)

Graphic techniques produced new and interesting images. Production studios developed more realistic set designs.

TELEVISION

Much of the theory for television broadcasting was known in the 1800s. It was not until the 1920s that the first working models were made. In 1923, Vladimir Zworykin invented the iconoscope and the kinescope. These devices were the basis for television.

In the 1940s advancements in television came quickly. In just a few years it became a very popular communication device. TV could transfer information more quickly than any other form of communication (other than talking face to face), Fig. 2-14.

Fig. 2-14. Over 75 million households in the U.S. have at least one television set. Television viewing is a popular activity. (Zenith Electronics Corp.)

COMPUTERS

Computers are devices that perform calculations and process data, Fig. 2-15. Several early computers were used before the automatic computer. The abacus was one. Adding machines with punch cards were another. The first practical computer was proposed by Charles Babbage as early as 1812.

Although Babbage's ideas were never used, the theory for such a machine existed. In 1944, Howard Aiken constructed the first mechanical computer, the Mark I. Electronic models followed shortly.

Fig. 2-15. Computers are used for many types of communication. Word and data processing machines are common in our modern information age. (A.B. Dick Company)

IMPROVING THE INVENTIONS

In 1948 scientists at Bell Laboratories invented the TRANSISTOR. This item either amplified or controlled electronic signals. It quickly replaced larger, more expensive vacuum tubes. As a result, transistors completely changed the electronics industry.

Science also extended itself into space technology. A global communication system was proposed in 1946. This became reality in 1960 when Echo I (a communication satellite) was launched. To date, many satellites have been placed in orbit, Fig. 2-16. Current satellites allow instant communications around the world. Telephone and TV signals are sent by satellites.

Advancements in microelectronics improved communication systems. This includes the integrating (combining into a whole) of many components that perform various functions. INTEGRATED CIRCUITS are fine examples, Fig. 2-17. These are complete systems manufactured on a single silicon chip. Electronic equipment is

Fig. 2-16. Communication satellites beam radio signals to all parts of the earth. Modern satellites orbit our planet far out in space. (Ford Aerospace & Communications Corp.)

Fig. 2-17. This electrical engineer is designing a new integrated circuit. The circuit is shown hundreds of times larger than actual size. When produced, it will be smaller than a fingernail. (Zenith Electronics Corp.)

reduced in size, needs less power, and produces less heat. The integrated circuit was invented in 1958.

The future of communication technology looks bright. Many changes lie just ahead. Picture telephones may become common in our homes. Foreign languages will be translated by computer. Satellites may permit direct home television reception. We truly live in an "information age."

SUMMARY

The Renaissance was when communication technology started to grow. Since then many new devices have been invented. The few mentioned in this chapter are landmarks. They paved the way for our modern systems.

KEY WORDS

All the following words have been used in this chapter. Do you know their meanings?

Amplification, Computer, Daguerreotype, Drafting, Hieroglyphics, Integrated circuit, Kinetoscope, Movable type, Photograph, Photography, Radio, Recorder, Satellites, Telegraph, Telephone, Television, and Transistor.

TEST YOUR KNOWLEDGE — Chapter 2

(Please do not write in this text. Place your answers on a separate sheet.)

1. The ability to make light, sound, and feeling more intense is called _____ .
2. When people first communicated with their voices, they used a system of spoken words. True or False?
3. What are hieroglyphics?
4. Briefly describe what happened during the Renaissance.

MATCHING QUESTIONS: Match the definition in the left-hand column with the correct term in the right-hand column.

5. __ A communication device that transfers information through audio and visual images.
6. __ Transmitted voices over wire.
7. __ System based on the practice of signaling others with flags.
8. __ Projection of moving images.
9. __ The "wireless" telegraph.
10. __ Used to preserve voices and other sounds.

a. Telegraph.
b. Telephone.
c. Radio.
d. Motion pictures.
e. Recorder.
f. Television.

11. What were the first practical photographs called? How did they get this name?
12. List three improvements of the first communication inventions.

ACTIVITIES

1. Visit a science and industry museum. Study displays that deal with early communication developments and inventions.
2. Make a list of communication devices found in your home. Decide what invention these devices came from. Bring your list to class and compare it with your classmates' lists.
3. Research an invention in communication technology that interests you.
4. Prepare a time line of the major developments of communication technology. Use drafting instruments such as a straightedge, T-square, and drafting pencil.

5. View films on the development of various technical devices in communication.
6. Have a representative of a telephone company speak to your class on the history of the telephone. What new functions can we expect to see in the future?
7. Use a microcomputer to make a list of major developments in communication technology.

Communication technology has made great progress in recent years. Experiments with laser light will provide growth in many areas. (Ford Aerospace & Communications Corp.)

3 | The Communication Process

The information given in this chapter will enable you to:
○ *Describe how messages are transmitted using a communication process model.*
○ *Explain the function of the each part of the communication process.*
○ *Explain the importance of accuracy in the communication process.*

You learned earlier that communication is a process of exchanging information using our senses. Information is sent from a source to a destination with codes and signals. Designing, coding, transmitting, receiving, and storage systems are needed for the transfer of communication. Feedback is also necessary, but often overlooked. Interference is unnecessary, but often present.

DESIGNING

The SOURCE is the starting point of messages to be sent. A source might be a machine, a person, or any supplemental system. A message is designed to be transmitted to others. The DESIGNING PROCESS includes ideation, purpose, and creation. IDEATION means getting an idea. Our skills, knowledge, and senses are then used to add purpose and create the message, Fig. 3-1.

IDEATION

Suppose you were going to be late for dinner. What would be running through your mind? You might feel hungry. You could be thinking about a cold meal. The development of this message involves sensing (feeling). An idea is created from these feelings. These thoughts were developed by the ideation process.

After the need to communicate is established, a message must be constructed. Previous ex-

Fig. 3-1. Many ideas and thoughts ''pop'' into our minds. We often remember happy thoughts or dream about future events.

perience is important to this process. A past situation or problem might come to mind. You may remember eating a cold dinner the last time you were late. Reference to earlier experience helps form the message. This is a reflective (thinking back) process, Fig. 3-2.

The message must now be further developed. Key questions still exist that need to be answered. These include who, what, where, when, and why. For example, WHO will be given the information? WHAT messages are to be exchanged? WHY will

Fig. 3-2. Does this photograph remind you of summer vacation? How you relax during vacation time is based on reflection of past events.
(Robert Maxham, Harte-Hanks Communications, Inc.)

one message be more effective than another message? Deciding WHERE and WHEN the communication process will occur is important. Like a good newspaper article, these questions are used to cover all the details.

The final phase of ideation is an evaluative process. That is, you must rely on skill and prior experience to transmit the message. How might the information be exchanged? Was the procedure successful when used before? Was a verbal or written message used? Perhaps a telephone call might be best. All of this depends on your skills and past experience. The means of developing messages is an important stage in communication.

PURPOSE

In designing a message, one must have a purpose. We often classify purposes in any of four groups. Messages can inform, instruct, persuade, or entertain, Fig. 3-3.

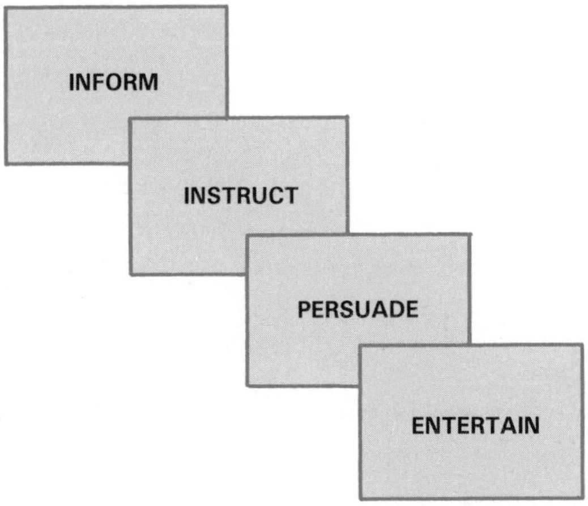

A message that informs tells a destination (person or machine) certain information. We read information in books and newspapers. The news programs on TV inform us of world news. Our parents also tell us important information.

A message that instructs provides direction. That is the purpose of this book. We will instruct (or teach) you how to use communication technology. Machines can also instruct. For example, a computer often receives instructions from other machines, Fig. 3-4. Various information is easily exchanged.

A message that persuades convinces someone to take action. Advertising messages are designed to persuade the public. They tell us about new and improved products, Fig. 3-5. We often buy these items because of these persuasive advertisements.

Finally, a message that entertains attempts to amuse people. Most television shows entertain us.

Fig. 3-4. This drawing of a manufactured part was first drawn on a computer screen. Then the computer "told" the printer to draw the item on paper. (Bausch & Lomb)

INFORM

INSTRUCT

PERSUADE

ENTERTAIN

Fig. 3-3. Purposes of messages.

Fig. 3-5. Many posters and signs feature messages intended to persuade others. How does this sign try to influence your thinking?

Other common media are designed for the purpose of entertainment. These include magazines, novels, and music.

CREATION

All messages include a thought process, or ideation, and purpose. The next stage is the actual creation (or design) of the message. CREATION is the assembling or recording of ideas. The idea is then arranged to fit our purpose. In speaking, this process occurs automatically in our minds. We think of a topic and talk about it with ease. What if you were asked about the weather? Your senses could create a description of current conditions.

However, we are concerned with the technical means of transmitting messages (communication technology). These messages usually require much planning. The text of a newspaper must be written and edited. In the same way, radio and TV scripts are carefully prepared. Each scene of a television show is written before production starts. Drafters also organize their thoughts on paper. They complete blueprints of items to be produced, Fig. 3-6. This is a form of prior planning.

All messages are generated in this manner. Authors and designers start their work in the ideation phase. They begin with sketches or notes on paper. Technically, we may call this an outline or DRAFT. Refinements will improve the message. Many rough sketches are necessary to select the best image. The early drafts of a television script will be refined many times. As writers prepare the story, various plots are considered. Good writers ask co-workers for advice, Fig. 3-7. The final copy (text) is the result of these efforts.

In summary, the creation process is quite complex. It includes thought, focusing, and creation. The acts of selection and revision follow. The final copy is now ready for use. However, early attempts may need refinement. Several drafts are often completed. Helpful suggestions from others are sought. The final design will be the best possible choice.

Fig. 3-7. Newspaper reporters often ask for advice from fellow workers. Do you ever ask classmates for helpful suggestions? (San Angelo Standard Times)

CODING

After a message is designed, it must be coded. CODES are vehicles for transmitting messages. This process differs greatly between various communication devices. Depending on the method used, we receive different types of codes. Common categories include signs, impulses, sound waves, and beams of light. These are typical systems of coding messages.

Sign codes are often handwritten. Drawings, paintings, and symbols are all common examples, Fig. 3-8. Many forms of communication use electrical impulses. Computers, telephones, and radios

Fig. 3-6. This set of house plans is being prepared for a construction firm. Drafting is a major part of technical communication. (Teledyne Post)

Fig. 3-8. Various types of signs and symbols warn us of hazardous areas. Do you recognize any of these coded messages?

all rely on these impulses. Some devices change impulses into sound. Radios and cassette players produce sound waves from electrical signals. On the other hand, bells and chimes operate by mechanical means. Sound is obviously an important communication channel. Light waves represent another system of coding. Television and motion pictures are perfect examples of light wave codes.

As was said, codes are vehicles for transmitting messages. A diagram of the coding techniques appears in Fig. 3-9. The source, or sender, must encode the message. This means the receiver must decode the information. Methods of decoding messages are covered later in this book.

TRANSMITTING

After a message has been designed and coded, the next step is TRANSMITTING, or sending, the message. If you write a note to a friend, how will it be transmitted? A piece of paper could be delivered in person. The note might be mailed in an envelope. Short messages may be transmitted by computers. Each of these examples is a way of transmitting information towards a destination. They are called MEDIUMS. Different mediums of transmission are available. Frequent channels include air, fluid, and solid material such as wire or gears.

You learned in Chapter 2 that radio signals travel through the atmosphere (air). These waves

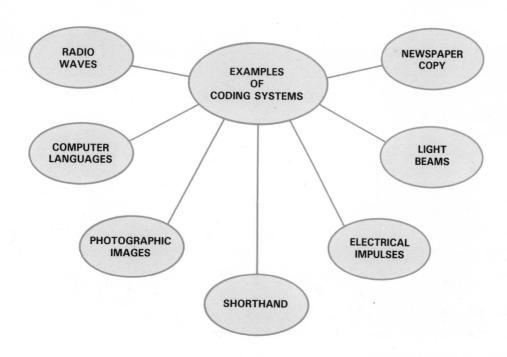

Fig. 3-9. Many types of codes are common in communication systems.

of electrical energy also carry TV broadcasts. The sounds created by whistles and bells are transmitted through the air, too. Other types of signals or electrical pulses travel along wires. Telephone conversations are transmitted through wires. Wires, signs, books, and gears are all examples of solids used in communication. Transmission of signals through fluids is not common to most of us. One example includes locating underwater objects using sonar equipment.

Not all messages are sent through a single transmitting medium. Television broadcasting is an example of using different mediums for communication. A televised message changes form many times before the public can enjoy it. Various signals are designed and transmitted. Verbal, written, electrical, and visual signals are used in the TV broadcasts. Different messages have been designed, coded, and transmitted. Thus TV programs demand the use of several human sensory systems.

RECEIVING

The communication of any message involves a destination. Someone or something must receive the information. They also need to decode (or interpret) the message. This is performed by the receiver. The exchanged information is then understood.

Decoding is important for success. It adds meaning to many communication efforts. For example, we are not able to see or hear radio waves. Fortunately, stereo receivers convert the signals to pleasant sounds.

The same process takes place when we read this book. Our knowledge of letters and words is important. We are able to understand the printing and pictures. We, as destinations, have our brain decode the signals. Prior skills helps us rebuild useful information from the reading.

STORING

The final step in the communication process is the storing of messages for later use. This procedure is known as STORAGE. There are many reasons for storing messages. Keeping important knowledge in books and recording news events for historical purposes are just two reasons.

Information is stored in many ways. Individuals often store information by memorizing details. We "teach" machines in much the same manner. Computers have an internal memory, Fig. 3-10, that is used for storage. Information is also recorded as written symbols. Books, magazines, and

Fig. 3-10. While the keyboard and screen of a computer are very small, large amounts of information can be stored in its electronic memory. (Sperry Corp.)

newspapers all contain recorded messages. Other methods for storing information exist. Mechanical, electronic, and various film mediums are common. Film storage includes photographs, slides, and microfilm. Microfilm stores much information in a small place. Mechanical recording was once quite popular. Some computers still operate with punched cards. Other industrial machines are controlled with programmed tapes, Fig. 3-11. However, electronic systems have replaced most of this equipment.

Storage of communicated work is important. The retrieval (recovery) of information for later use is a necessary process. Methods of storing information are outlined in Fig. 3-12.

FEEDBACK

The process of communication is not yet complete. An additional element is feedback. Successful communication is often the result of feedback to the sender.

FEEDBACK is a sign that a message has been

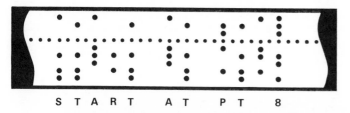

S T A R T   A T   P T   8

Fig. 3-11. Paper tape is used to guide many manufacturing activities. Holes punched in the tape provide "directions" to the machine.

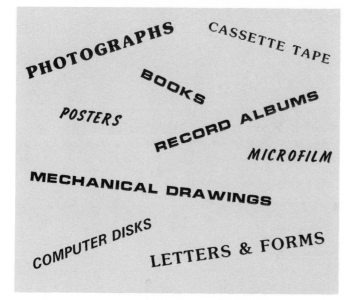

Fig. 3-12. How many of these common types of "storage systems" do you recognize?

received. This usually occurs by exchanging a signal. For example, suppose a person sends a message to a friend. They would like some type of feedback from that person. Did the message arrive? Was it understood? Many questions could arise after sending the message. The key is that some type of response is communicated.

Feedback may be direct or indirect. DIRECT FEEDBACK involves spoken or written words. It may also include gestures or body movements. In any case, direct observation of the receiver is possible. INDIRECT FEEDBACK results from observing later actions. These actions will prove that the message has arrived.

INTERFERENCE

Interference is a distortion of signals intended for the receiver. Interference is caused by the reception of undesired signals, Fig. 3-13. It is often called NOISE.

Most interference is caused by talking, daydreaming, etc. This noise interferes with the communication of the message. Mechanical interference is caused by failures in equipment. A broken television set is a common example. Electrical interference is very similar. The static your television picks up during a thunderstorm is a common form of electrical interference. Interruptions in radio signals are also considered noise.

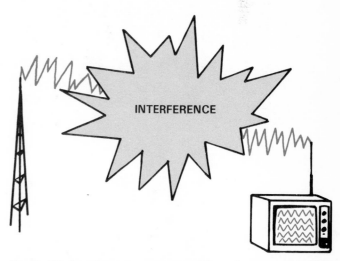

Fig. 3-13. Interference causes problems in many types of communication. Static on the radio is a common disruption.

ACCURATE COMMUNICATION

We use communication methods today without much thought. This is because of the numerous means available to us. However, what value are they if they are not used correctly? How important is an accurate message?

What if you need to get a message to a friend? Will they receive something other than what you wanted communicated? If only verbal exchanges are used, the message could be misunderstood. What would happen then?

Careful planning may prevent inaccurate transmissions of information. For example, do you think it requires accurate communication to play football? The players must know what to do on each play, Fig. 3-14. Who will carry the ball? What is each individual's assignment during the play? When will the ball be snapped? Where will each player run? How does each player contribute to the success of the play? Also, coaches may ask why the play worked or failed.

These are key questions if the team hopes to succeed. They also bring to light some essentials of accuracy. You learned earlier in this chapter that in communication we should always consider the who, what, when, where, how, and why of the message.

We require an increasing amount of communication. Information is exchanged among people, machines, and governments. Business and industrial groups are included in this trend. Complex new laws and systems of ownership demand improved methods of communication. Errors and misunderstandings may occur if people do not communicate clearly. The questions who, what, when, where, how, and why must be addressed. Only then will accurate communication be achieved.

Let's examine a situation where exact communication is important. Suppose your job involves ordering materials for a firm in California. You receive a requisition (order) from your boss. The company needs 10,000 special plastic fasteners. These materials must be obtained from a company in Ohio within two weeks. What must you do to purchase the fasteners?

First, a material order form should be completed accurately, Fig. 3-15. The department that needs the fasteners will be identified (who). A description of the type of fasteners needed is important (what). The form will also include the date the items are needed (when). Fourth, the company in Ohio must be located. The address in your files tells you the location for placing the order (where). The method for placing the order is dependent upon time constraints (how). The mail might be too slow so a telephone call is necessary. This all brings up the question, "Why?" The order is being placed to meet production demands. Failure to complete this communication process will lead

Fig. 3-14. What happens if the quarterback gives his teammates the wrong play? Poor communication among the players might result in a lost game.

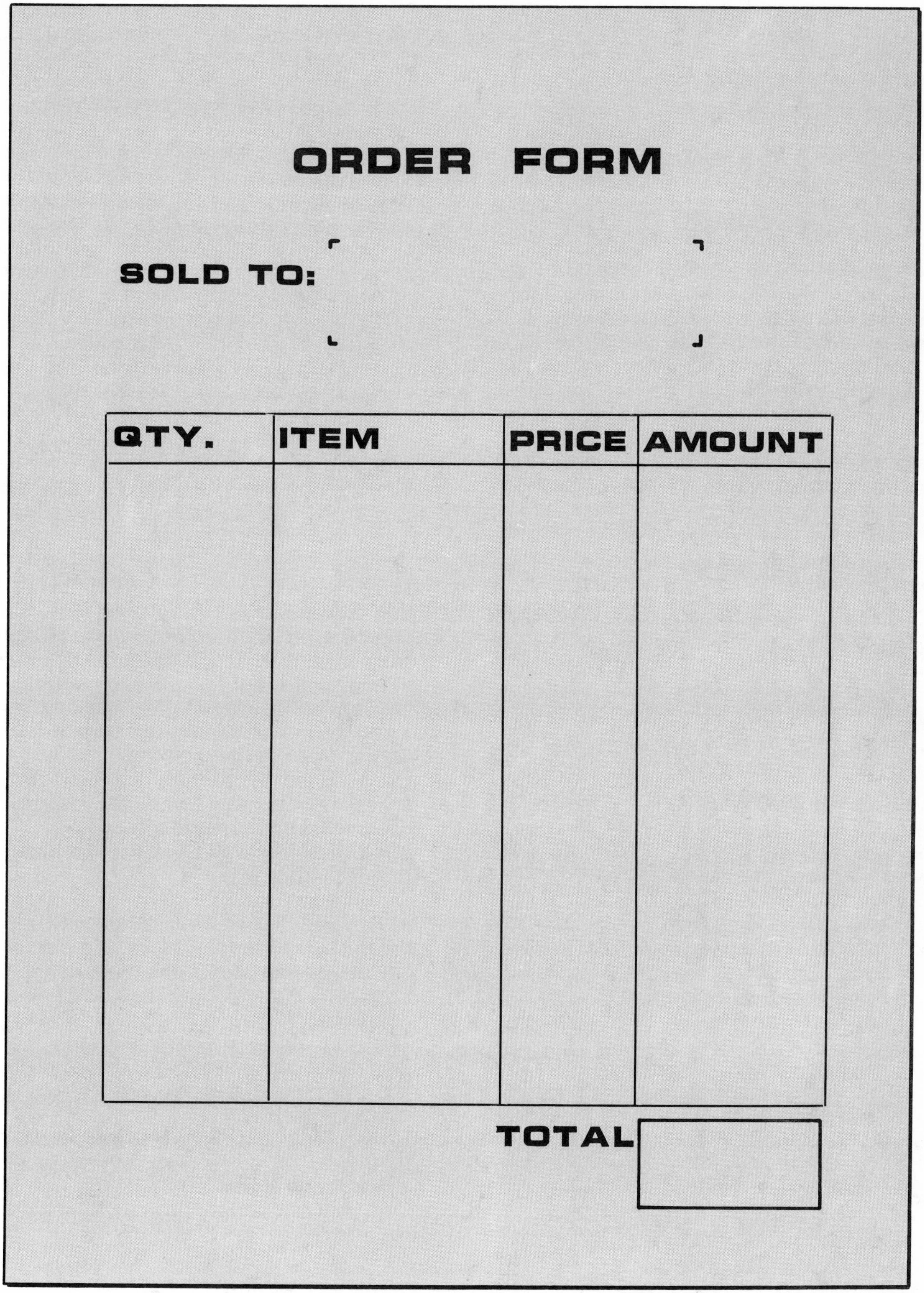

ORDER FORM

SOLD TO:

QTY.	ITEM	PRICE	AMOUNT

TOTAL

Fig. 3-15. Order forms represent a common type of communication. Businesses use many tons of paper each year in exchanging information.

to delays. Or worse, the wrong items may be delivered.

By now you see the effects of poor or inaccurate communication. It is the cause of numerous mistakes. This leads to confusion, wasteful use of resources, and lost time. Accuracy in communication technology is important.

SUMMARY

This chapter increases your knowledge of the communication process. The steps outlined here are the design, coding, transmitting, receiving, and storing of messages. The influences of feedback and interference were reviewed. Four purposes of communication have been identified. These purposes include informing, instructing, persuading, or entertaining. Finally, the accuracy of communication was discussed. Questions such as who, what, when, where, why, and how are important to accuracy.

KEY WORDS

All the following words have been used in this chapter. Do you know their meanings?

Accuracy, Codes, Coding, Creation, Direct feedback, Designing, Feedback, Ideation, Indirect feedback, Interference, Noise, Purpose, Receiving, Source, Storage, Storing, and Transmitting.

TEST YOUR KNOWLEDGE — Chapter 3

(Please do not write in this text. Place your answers on a separate sheet.)
1. Information is transmitted from a _____ to a _____ with _____ and _____ .
2. The source is the starting point for a message. True or False?
3. The design process includes:
   a. Ideation, purpose, source.
   b. Coding, transmitting, receiving.
   c. Ideation, purpose, creation.
   d. Instruction, information, persuasion, entertainment.

MATCHING QUESTIONS: Match the definition in the left-hand column with the correct term in the right-hand column.

4. __ The process of making up a message.
5. __ The process of putting the message into a certain form for transmitting.
6. __ The process of sending the message to a destination.
7. __ The process of saving information for later use.
8. __ The process of securing and decoding a message.
9. __ The sign that a message has been received.

   a. Designing.
   b. Coding.
   c. Transmitting.
   d. Receiving.
   e. Storing.
   f. Feedback.

10. What is interference?
11. Why is accurate communication so important?

ACTIVITIES

1. Make a drawing to show how the communication process works.
2. Invite a writer or illustrator to class. Ask your guest about the processes he/she goes through in designing a story or piece of artwork.
3. Make a list of transmitting mediums (air, wires, etc.) with examples of communication devices that use those mediums.
4. Invite a radio engineer or disc jockey to visit your class and explain how radio messages are transmitted and received.
5. List different ways messages have been stored throughout history.
6. Design a message. Whisper your message to a classmate. Then have each student whisper the message to the student next to him/her. Let the message go all the way around the classroom. How does this show accuracy and interference?
7. Act out different examples of "body language." Try to guess what each student is "saying." How do these actions show feedback in communication?
8. Invite a newspaper reporter to visit your class to discuss the importance of accuracy in writing news stories.

4 Social and Cultural Influences

The information given in this chapter will enable you to:
- Describe the purposes of communication.
- Cite uses of communication technology in modern society.
- Identify social issues associated with communicaton technology.

Modern society relies heavily on communication systems. We use them in our personal lives, at work and play. For this reason, this era is called the "information age." This chapter will focus on how we are affected by modern communication technology.

Look around your home. Newspapers, television, smoke alarms, computer games, books, and magazines are just a few of the communication devices present. Each influences our lives in a different way. Each has an impact on our attitudes, feelings, and knowledge.

PURPOSES OF COMMUNICATION

No wonder this period in history is referred to as an information age. Our daily lives depend on many communication devices. We awake to an alarm clock. Morning newspapers and TV shows entertain and inform us. The radio announcer tells us the weather forecast. Clocks tell us its time for school or work.

A typical school could not exist without communication aids. Bells signal the start of classes. Announcements are delivered over the PA system. Teachers rely on chalkboards, movies, and instructional recordings, Fig. 4-1. Books contain the knowledge we seek to complete assignments. Conversations with friends or teachers highlight the day. Even computers and calculators have helped us with school work, Fig. 4-2.

A trip down any street reveals other communica-tion devices. Road signs provide direction, Fig. 4-3. Traffic lights control the flow of cars. Store fronts describe products or services. Billboards advertise our favorite consumer goods. Telephone booths and fire alarm boxes line the streets, Fig. 4-4. Can you identify other useful communication devices?

Our journey might continue until we arrive home. The impact of communication technology is reflected here, too. The mailbox contains personal letters or magazines. Inside the home are more devices for communication. A telephone answering machine might have recorded our calls.

Fig. 4-1. Teachers use many communication devices to help students complete class assignments. (Bell & Howell)

Fig. 4-2. Calculators and personal computers are two popular examples of communication aids.

Fig. 4-3. Signs along roads or streets provide valuable information. These visual displays tell us simple, yet important, details.

Fig. 4-4. These items can both be used to notify emergency personal quickly.

Most homes have several TV sets and a stereo system, Fig. 4-5. We also enjoy video game sets and personal computers. Books, typewriters, and clocks are found in many homes, too.

The purposes of communication were briefly described in the last chapter. Most communication is meant to inform, entertain, persuade, or educate. We will explain them further now.

INFORMATION

To inform is to let people know what is happening. Many messages are intended to spread information. They inform others of interesting or important facts. For example, radio announcers inform us of special events. Newspapers carry news, weather, and sports stories. Have you ever read the graphics on a breakfast cereal box? The pictures and words on most packages are informative in nature, Fig. 4-6. Countless efforts are available to inform us of key messages.

In these instances, the media acts as an information device. It provides knowledge about facts or events. This knowledge often helps you plan daily affairs. It also aids in making decisions. People need and enjoy receiving information. This is a basic aspect of communication technology.

ENTERTAINMENT

The mass media is used basically as entertainment. When you watch a television show you usually want to be entertained. If you listen to the radio, you want to hear music you like. Many people also use mass media as a way to relax. You may know someone who likes to relax by reading

Fig. 4-6. The labels on most packages give the size, weight, and ingredients. What information is shown on these packages?

Fig. 4-5. Almost every home in this country has at least one television set and radio receiver. These communication devices keep us entertained and informed. (Zenith Electronics Corp.)

Fig. 4-7. How often do you watch sporting events on television? This also is a form of entertainment. (Zenith Electronics Corp.)

the newspaper before eating dinner.

Television is the most popular form of entertainment. Today the average household includes several television sets, Fig. 4-7. Video cassette recorders (VCRs) and computers are also common home entertainment devices.

EDUCATION

Communication technology is also used to educate. When we think of education, we usually think of schools. However, how do we learn in school? Most teachers tell us of valuable information. Speech is communication. Others show us how to do things. Action is communication. We may also learn by reading. These are all the result of communication technology. Instructional communication has become very popular today. Films, slide/tape shows, and educational television are several examples.

Today we have thousands of magazines, newspapers, and books to choose from, Fig. 4-8. Thousands of new books and articles appear every year. Many are instructional in nature. Home fix-it books show us how to sew or take photographs. While these books explain fun and relaxing activities, their major purpose is to educate.

Fig. 4-9. Many communication materials are used to persuade. What reaction do you have to this visual display?

Fig. 4-8. Magazines provide hours of enjoyment for their readers. Do you subscribe to any journals, magazines, or newspapers?

PERSUASION

Communication technology is used to influence people's decisions. Signs and banners urge us to make purchases, to vote, and to support certain issues, Fig. 4-9. Many advertisements attempt to persuade the way we act. Which type of soft drink do you buy? Which fast food restaurants do you enjoy? Many times our decision to buy a certain product often results from persuasive efforts.

Forms of persuasion differ greatly. For example, news events are reported in different ways. A particular issue might be important in one city but another city may totally ignore it. What you are shown or told influences your outlook on life.

USES OF COMMUNICATION

We have examined the four purposes of communication. Now we will examine the uses of communication systems. Two broad classifications can be easily identified. Humans generally communicate with other individuals or with large groups of people, Fig. 4-10. When large numbers of people are involved, we call it mass communication. Much of modern communication is transmitted by our mass media.

INDIVIDUAL COMMUNICATION

Both people and business use individual (one-to-one) communication systems. Personal notes and letters are simple examples. Using a calculator to check figures is another. Telephone conversations with friends are also individual uses. The public does not have access to these individual systems.

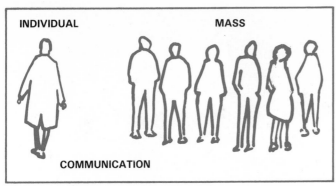

Fig. 4-10. The exchange of information always involves receivers, or an ''audience.'' An audience can be a single person (called individual communication) or large groups of people (called mass communication).

In business, individual use of communication is very important. The two most popular uses include the telephone and the postal service. Much information is transmitted by technical systems, Fig. 4-11. This is how business manages daily affairs.

The computer is also an individual use of communication. Business and private citizens all use computers. Most home systems are fairly small. Business systems, however, are usually very large. Information is kept in electronic files for future use. These files are called DATA BANKS. With these data banks, information is available for quick retrieval. This information may appear on a screen or be printed on paper, Fig. 4-12.

Fig. 4-11. Many people use banking machines to make deposits or withdrawals. These machines send messages (information) to the person's bank. These machines are often used during hours when the bank is closed.

MASS COMMUNICATION

Each day we receive messages aimed at many people. Each of us is part of a large audience for advertisers, the government, and other groups. Mass communication systems are used to provide information to large groups. The audience acts as the receiver in this communication process.

Fig. 4-12. Great amounts of information can be stored in data banks. The information needed can be typed (background) or photocopied (foreground).

Communication to large audiences relies on modern technologies. Television, radio, newspapers, motion pictures, and books are examples. Other mediums include billboards, road signs, and CB radios.

The impact of mass communication is important in modern society. None is more powerful than television. This medium provides a constant flow of messages. We see advertisements for toothpaste, autos, and food. News of local and world affairs is shown.

Radio is a key mass communication device. We hear our favorite music on various channels. Weather and news reports are broadcast many times each hour. Advertisements promote products, services, and special events. Again, the focus is to communicate with the public.

Mass communication technology is common in modern society. However, the trend is to make communication more personal. This alternative tries to focus on a particular segment of people. Messages are directed to particular groups. This is called TARGETING an audience, Fig. 4-13. For example, sporting goods companies will advertise heavily during football game broadcasts. These commercials are targeted at the sportsminded audience.

COMMUNICATION INFLUENCES

The influences of communication on society are often positive. We are entertained, informed, and

RECREATIONAL EQUIPMENT

RECORD ALBUMS

BOOKS AND MAGAZINES

CANDY BARS

FOOD AND DRINKS

Fig. 4-13. Advertisers create promotions for products that teenagers often purchase. Have you seen advertisements for any of these products?

Fig. 4-14. Closed-caption symbols are used in many television program guides to identify shows using subtitles for the hearing impaired.

educated through the use of various media. News and weather reports are an example of receiving useful information. However, communication technology can often have a negative influence on society, also. The distribution of false or misleading information is an example.

We will review several of the positive and negative influences communication has on society in this section.

INFORMATION IN SOCIETY

We have numerous communication devices available to us. We gain much from our communication materials (books, magazines, radios, televisions, newspapers, etc.). Because of this we have become the most informed country on earth. People in some countries rarely have access to so much information.

EDUCATION IN SOCIETY

Closely related to being informed, we are very well-educated. Our country guarantees education for all people. Citizens must remain in school until age 16. A large percentage of our population attends college. Education is available because of communication technology. Textbooks contribute to these instructional activities. Educators also use films, television, and audio-visual equipment. In fact, our entire society benefits from educational efforts.

FREEDOM OF SPEECH

Newspapers, magazines, television and radio stations, and all citizens depend on freedom of speech. Because of it we can express our opinion. For example, a newspaper can endorse (recommend) a certain person for a political office. And the evening news show can endorse another candidate for that same political office. Citizens have the right to carry signs and banners in support of a third candidate for the same political office.

ECONOMICS

Many people are employed in various communication activities. Writers, printers, and broadcasters create messages. Countless technicians and engineers work in the communication field.

All these people get paid for their work. They use their money to maintain a particular way of life for themselves and their families. Food, clothing, and various services are bought with this money. This money goes back into the economy and helps other people.

THE HANDICAPPED IN SOCIETY

Communication also helps in assisting the handicapped. The deaf can be informed or entertained by television shows with subtitles. These close-captioned broadcasts provide a written script on the screen, Fig. 4-14. Braille books and magazines help blind people to enjoy literature.

CLUTTER

Overexposure is a negative part of the mass media. We are often exposed to too much information. A term for this is INFORMATION OVERLOAD. Our minds just are not accustomed to processing so many messages.

Sources of clutter are all too common, Fig. 4-15.

Fig. 4-15. People often complain about too many signs along roadways. Is this an attractive sight?

We see numerous billboards and signs along roadsides. Television commercials suffer from this problem also. We see too many advertisements to clearly remember a select few.

LAWS AND REGULATIONS

With the growth of communication, federal laws and regulations were created to control various media. These rules limit many communication efforts. However, this regulation is designed to protect us. Various laws insure that the public is being served fairly.

The Federal Communications Commission (FCC) regulates radio, TV, and telephone communication. Other governmental agencies control the publishing industry. The results of these regulations are quite evident. For example, cigarette advertisements are prohibited on television. Also, obscene language is restricted in most media. And advertisers may not show alcohol being consumed on television commercials.

PROPAGANDA

The most harmful use of communication technology on society is propaganda. PROPAGANDA is the use of false or misleading information. It is used to force others to accept certain opinions. This is a negative use of communication methods.

SUMMARY

Our lives rely heavily on communication systems. From the beginning to the end of the day, we used communication devices. In the morning we wake up to alarm clocks. In the evening we relax by watching television or reading a magazine. These communication systems are used as means of education, entertainment, persuasion, or information. They can be used to communicate with individuals or with large groups. When large groups are involved, it is called mass communication. However, no matter what way communication systems are used, they do have influences on society. These influences can be either positive or negative.

KEY WORDS

All the following terms have been used in this chapter. Do you know their meanings?

Data banks, Education, Entertainment, FCC, Individual communication, Influences, Information, Information age, Information overload,

Mass communication, Persuasion, Propaganda, Regulation, and Targeting.

TEST YOUR KNOWLEDGE — Chapter 4

(Please do not write in the text. Place your answers on a separate sheet.)

1. Why is this era called the information age?
2. Which of the following is NOT a purpose of communication technology?
   a. Freedom of speech.
   b. Persuasion.
   c. Education.
   d. Information.
   e. Entertainment.
3. When large numbers of people are involved in the exchange of information it is called _____ .

Identify the following devices as individual or mass communication uses of communicating.

4. Radio.   _____
5. Letters.   _____
6. Postal service.   _____
7. Newspapers.   _____
8. Calculators.   _____
9. _____ is directing a message to a particular group.
10. Is information overload a positive or negative influence? Why?

11. The _____ regulates radio, television, and telephone communication.

ACTIVITIES

1. Conduct a school-wide survey to learn the number of telephones, televisions, and radios found in each student's home.
2. Take the TV section of the newspaper and review the shows for different times of the day (morning, daytime, 4-7 p.m., and 7-11 p.m.). Classify the programs as informational, entertainment, educational, or persuasive.
3. Listen to the radio between 4 and 7 p.m. on a weekday. List the communication purposes, with examples heard during the broadcasts.
4. Analyze the sections in several magazines to identify the parts that inform, educate, entertain, and persuade.
5. List the forms of communication technology you use in one day. Divide the list into individual and mass media communication uses.
6. Have a guest speaker from an organization for the deaf demonstrate sign language and other communication techniques used to assist the deaf.
7. Collect magazine and newspaper advertisements that inform, educate, entertain, or persuade. Classify these by purpose.

5 | Industries and Careers in Communications

The information given in this chapter will enable you to:
○ *Identify the major communication industries.*
○ *Understand the importance of information industries in modern society.*
○ *Describe the products and services offered by different communication industries.*
○ *Recognize the variety of career opportunities in the field of communication.*

Creating and transferring information is one of the largest businesses in the world. These business activities are collectively called the COMMUNICATION INDUSTRY. Millions of individuals work in various jobs involving the transfer of messages. Some people produce and broadcast television shows. Others write, illustrate, and publish newspapers and magazines. Computer programmers and photographers are also involved in the communication of information. Examples of activities in this field are almost endless. But all activities in the communication industry have one thing in common: they are designed to make a profit (money) by spreading information to others.

The size and complexity of the communication industry is astonishing. In this country alone, 60 million (60,000,000) newspapers are circulated each week. Over 120 billion (120,000,000,000) pieces of mail are delivered yearly. Our home telephone is connected to some 500 million (500,000,000) phones around the globe. Some of the larger communication companies often employ thousands of workers. Several other "unusual" communication systems are shown in Fig. 5-1.

In this chapter, we will examine several business activities associated with the communication

Fig. 5-1. Two types of communication systems. A—Store front signs. B—Satellite dish.

industry. We will focus on those industries that use technical systems to transfer information. We will examine how typical industries are organized and operated. In addition, we will explore the many challenging professions available in the communication field, Fig. 5-2.

COMMUNICATION INDUSTRIES

The economic activities in our country are mainly conducted by business enterprises. We call these organizations COMPANIES. The term INDUSTRY best describes a group of related businesses. For example, radio and television stations are classified as parts of the broadcasting industry. The printers of books, magazines, and other graphic materials are part of the publishing industry, Fig. 5-3.

Very simply, industries add or create value in products and services. Resources such as materials,

knowledge, and money are used to create this value. Individuals and businesses then pay the industries for these products and services. Industries hope to generate a profit (money) in this manner, Fig. 5-4.

Eleven types of communication industries have been identified in Fig. 5-5. For our discussion, we will examine each of these industries more closely.

Customers buy . . . .

Products & Services

Fig. 5-4. Companies charge others for the products and services that they offer.

Fig. 5-2. Railroad dispatchers are an important link in this communications field. They must warn trains of conditions on the tracks ahead. (Santa Fe Railway)

COMMUNICATION INDUSTRIES
Commercial Art and Design
Engineering Drafting and Design
Printing/Publishing
Photography
Advertising
Broadcasting
Recording
Cinematography
Computer/Data Processing
Telecommunications
Service

Fig. 5-5. The 11 communication industries.

Fig. 5-3. Small printing shops turn out the majority of printed materials in this country. (AFT—Davidson)

COMMERCIAL ART AND DESIGN

Many companies specialize in the creative design of commercial items. These businesses are called DESIGN FIRMS or DESIGN AGENCIES. Design firms range in size from one person to hundreds of people. The work of design firms is visible in many products, Fig. 5-6.

Commercial designers are very talented people. They have developed their skills through many years of work. Many designers have attended schools and colleges to study art and design. As a result, they are very good at sketching and drawing. This permits them to put ideas on paper for others to evaluate and enjoy, Fig. 5-7.

The work of designers is everywhere in our lives, Fig. 5-8. They design common items like posters, greeting cards, and candy wrappers. Other examples include postage stamps and record album jackets. Products like these are carefully designed for your use and/or enjoyment.

Fig. 5-8. We use many materials produced by creative artists and writers.

Fig. 5-6. Designers create many of the labels, signs, and other visual displays we see everyday.

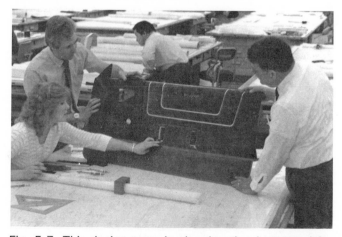

Fig. 5-7. This design team is planning the door panel for a new automobile. Their ideas started out on paper, in the form of engineering drawings (shown on the table). (General Motors Corp.)

Large agencies often employ hundreds of individuals. Each person specializes in a particular area. Layout artists develop plans based upon the desires of a client (customer). Staff artists are hired to produce the detailed sketches and drawings. Printers reproduce the designs on paper or other stock. During the entire design process, a project director supervises the job. This person coordinates the work of artists, photographers, printers, and many others.

ENGINEERING DRAFTING AND DESIGN

Sketches and drawings are excellent ways to communicate ideas. Presentation of illustrations is called DRAFTING, Fig. 5-9. Drafters use symbols and shapes to develop designs for many objects. All products and structures are carefully planned before being produced, Fig. 5-10.

Drafters in industry develop plans for cars, toys, furniture, and other products. Their designs are often called ENGINEERING DRAWINGS. Other drafters specialize in planning homes and buildings. We call these individuals ARCHITECTURAL DRAFTERS (or architects). Your school building was designed by an architect. Many drawings were required to completely explain how the building would be built. When roadways and large buildings are planned, thousands of drawings are needed.

Fig. 5-9. This drafter is using a computer to complete a drawing. Computers help speed up the design process in many industries. (Computervision Corp.)

Fig. 5-11. Drafters are able to complete many forms of drawings. Examples of this work are displayed in the background. (General Motors Corp.)

Fig. 5-10. The design of this support is being carefully examined on the computer screen. (Bausch & Lomb)

Fig. 5-12. Notice how the technical illustrations of these video recorders (background) match the actual products (foreground). (RCA)

Drafters must know how to produce many types of drawings, Fig. 5-11. For example, plans for electrical devices are called SCHEMATIC DRAWINGS. Another form of drawing is a TECHNICAL ILLUSTRATION. This is a finely detailed drawing that resembles a photograph, Fig. 5-12.

Drafters are usually skilled workers. They attend special schools for training. Much of their education includes taking art, math, and technical courses (physics, drafting, etc.). Most will also work as apprentices or detail specialists for several years in design firms. Some areas of drafting require this experience before taking certification tests. Architectural drafters must pass these tests to become fully-licensed architects. Civil and product engineers must pass a similar exam.

Modern drafting is often done completely with the aid of computer systems, Fig. 5-13. Computers allow the drafter to develop blueprints and detail drawings much faster. If a certain shape or symbol is needed, the designer can "draw" the item by pushing a key. Time spent in the design and drafting process is reduced with the aid of computer systems.

PRINTING/PUBLISHING

Printing and publishing businesses are an important part of the graphic arts industry, Fig. 5-14. This group of related businesses make money by printing various materials to be used by others.

Printing firms are classified by the type of printing service they offer. For example, commercial printers do small jobs, involving limited numbers of copies. This would include business cards, stationery, and flyers. These QUICK PRINTERS usually provide "while you wait" service. They take advantage of electrostatic copiers to offer fast service to customers, Fig. 5-15. Another group of printing firms produce legal documents and related materials. We call these shops, SPECIAL PURPOSE PRINTERS. Finally, the most familiar printing companies are called PUBLISHING HOUSES. They print newspapers, magazines, and many forms of books.

Larger publishing houses often hire hundreds of workers. Separate employees help write, edit,

Fig. 5-15. This is the type of copier you may find in a quick print shop. This particular copier is able to make two-sided copies from a one- or two-sided original. (Eastman Kodak Company)

Fig. 5-13. Large computers are often used to produce engineering drawings in modern industry. (Computervision Corp.)

and illustrate articles or stories. Other workers are in charge of production tasks such as layout and printing. Typesetters transform all text and captions into the actual type to be used in the book, magazine, etc.

Obviously, the training or education for each employee is different. Writers and editors must know how to prepare feature articles, Fig. 5-16. They usually study journalism or English in college. Talented artists use their training in design to develop the required drawings and artwork.

Fig. 5-14. Newspapers are a common product of the printing or publishing industry. Workers are shown preparing the daily edition for delivery. (San Angelo Standard Times)

Fig. 5-16. This newspaper writer obtains information over the phone in order to prepare an article for the daily edition of the paper. (Ball State University)

Typesetters' keyboard skills have been improved through practice and on-the-job experience.

Production workers in publishing houses require specialized training, also. Press operators are responsible for the actual printing of all materials, Fig. 5-17. The printing presses these operators work with are very complex machines. Therefore, each operator often has had technical schooling, in addition to on-the-job training.

Other production tasks include platemaking, process camera work, bindery work, and paper cutting. Workers in these areas usually attend college or technical schools to learn their trade. Many times they are represented by a trade (labor) union. As union employees, they will often spend several years learning press techniques from people with many years of experience. These experienced people are called MASTER PRINTERS. The time spent observing and working with a master printer is called an APPRENTICESHIP.

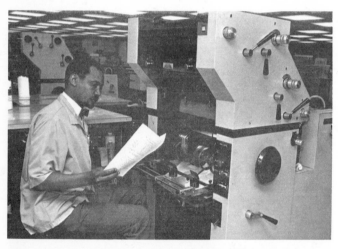

Fig. 5-17. This press operator checks the quality of a reproduction during the printing of a job. (AFT—Davidson)

PHOTOGRAPHY

Many people enjoy taking pictures simply for fun or relaxation. Professional photographers, however, also earn money for the pictures they take. These photographers are employed by others for commercial purposes. For example, newspapers hire press photographers to "shoot" pictures of news and sporting events, Fig. 5-18. Most television stations maintain a large photographic staff for the same reason. In contrast, certain individuals prefer to work for themselves. We call these people FREELANCE PHOTOGRAPHERS. They often take pictures for small events such as weddings, family reunions, or school yearbooks.

Fig. 5-18. Press photographers crowd the sidelines to get the best pictures for their newspaper. (John Metzger)

A career in photography demands hard work and long hours. In addition, photographers must have training in many areas. Most photographers are knowledgeable about camera equipment, lighting, and visual composition. Colleges and technical schools offer many courses in these areas. In addition, photographers improve their skills through hours of instruction and practice.

ADVERTISING

We see many forms of advertisements every day. Television commercials and magazine ads are common examples of advertising. Other media include billboards and banners, Fig. 5-19. All advertising work involves communication between companies (or organizations) and the public.

Most forms of marketing rely on communication technology to help "spread the word." The mass media (TV, newspapers, radio) is an effective way to inform and persuade large groups of people. After all, nearly everyone listens to the radio and watches television. Using these two mediums for advertising purposes is very effective. Other systems are equally as effective for advertising purposes. Various graphic advertising

Fig. 5-19. How many forms of advertising do you see on this street?

mediums are shown in Fig. 5-20.

Individuals in the advertising profession are highly trained. Their knowledge of design, marketing, and public relations is important. Therefore, advanced schooling is critical to becoming successful in this field. Classes in art, design, and production are usually required. In addition, courses in business are important, too. Advertising professionals then improve their talents through on-the-job experience.

Many companies do not have the time or resources required to develop an advertising program. These companies often hire others to design their promotional materials. A large organization that specializes in this task is called an ADVERTISING AGENCY. Most advertising agencies employ dozens of artists, writers, and related personnel. Each person has a strong background in one area of advertising work. The agency offers the creative services of their employees to outside firms (for a fee). Major advertising campaigns require teamwork among the agency staff.

The work of advertisers is fun and rewarding. Using creative talents to develop an ad is exciting and challenging. However, preparing a successful advertising program is hard work. Whether it's designing a billboard or recording a radio ad, much effort is expected of advertising personnel.

BROADCASTING

The broadcasting industry includes work in the fields of radio and television. There are several hundred TV stations and nearly 8,000 radio stations in this country. Most local stations are representatives of major networks such as NBC, ABC, and CBS. Independent and public stations function at a smaller scale.

Typical television stations are complex organiza-

tions. It takes a large staff to prepare and produce a TV show, Fig. 5-21. Programs must be developed by writers and reporters. Advertising time (commercials) must be sold by the marketing department. The production staff operates and maintains studio equipment, Fig. 5-22. Business and personnel matters are handled by company officials.

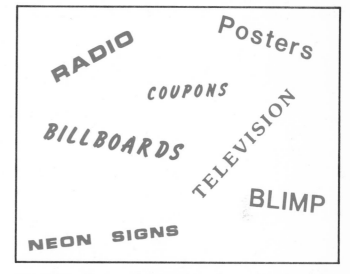
Fig. 5-20. These are examples of several advertising mediums.

The structure of a radio station is much the same as a television station. Feature programs and news reports must be prepared for broadcast. Electrical engineers keep the transmitting equipment in operation. Radio disc jockies (DJs) read announcements, conduct interviews, and introduce musical numbers, Fig. 5-23. In fact, the success of a typical station will depend upon the popularity of their DJs. The better announcers attract a larger share of the market (more listeners). This leads to higher ratings and, generally, a more profitable station.

Radio and television broadcasting is a very challenging profession. Most positions in the broadcasting industry are "behind the scenes." Support personnel (like writers or directors) are rarely seen or heard on screen. Their jobs are far less visible than the newscaster or reporter. However, their work is still very important and rewarding.

Most people in the broadcasting industry have some type of advanced training for the positions they hold. For example, talk show hosts often have college degrees in communication. Directors and producers often hold degrees in mass communication or broadcasting. Obviously, studio technicians

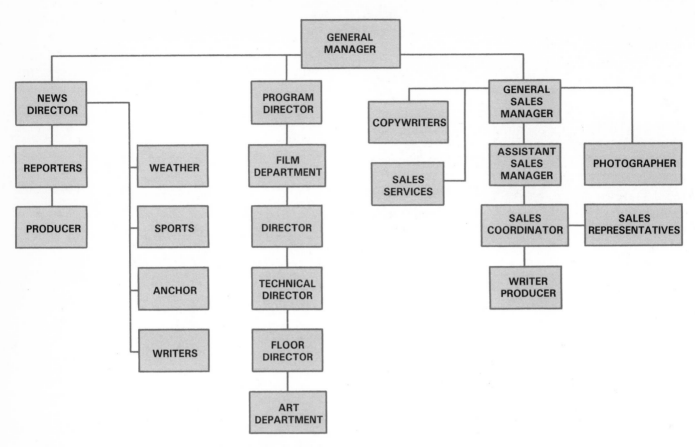

Fig. 5-21. Many people are involved in the organization of a typical TV station.

Fig. 5-22. The production staff is getting cameras and lights ready for the next broadcast. (Ball State University)

Fig. 5-23. A radio disc jockey introduces songs and reads announcements on the air. (Ball State University)

attend technical or trade schools to learn about electronic equipment and systems. Almost everyone in this field will start their career in a small, local station. The brightest and most talented individuals, however, will soon discover new opportunities in larger stations.

RECORDING

The recording business is a $2 billion industry in our country. Money is spent to see popular recording stars at concerts. Albums and cassettes of rock, country, and other music are bought often, Fig. 5-24. Radio stations play our favorite hit songs. This describes just a part of the recording industry.

Fig. 5-25. Sound technicians check the audio level during production sessions. (Ball State University)

Fig. 5-24. Albums and cassettes are two of the more common products of the recording industry.

Individuals in the recording industry have varied backgrounds and skills. Most studio and recording personnel are highly trained. They must understand both acoustics (the science of sound) and electronic equipment. Microphones and instruments must be arranged correctly to obtain the perfect sound quality. Recording sessions for a demonstration or master tape often involve long hours in the studio, Fig. 5-25.

CINEMATOGRAPHY

Cinematography is the science of motion-picture photography. This industry is more commonly known as the motion-picture industry. Like the recording industry, the motion-picture industry is a very popular and profitable form of communication, Fig. 5-26.

Making a motion picture is a complex task. Many talented people are needed to write, film, and edit the movie. Casting agents select actors and actresses for each character. Set designers see

Fig. 5-26. The motion-picture industry entertains millions of people every year.

that the stage (or scene) and props are built. Rehearsal and production time is scheduled to complete the filming work. The camera crew and their support team film all action. After all the camera work is completed, editors "cut" the film to final length and content. The sound track (voices and music) is then added by studio technicians. Various directors and managers supervise the entire production. Finally, the distribution process begins. Movie theaters compete to obtain the most popular movies available. Distribution companies contract and ship motion pictures to theater owners. Finally, the film is shown in your local theater.

Careers in the motion-picture industry require a great deal of hard work. When filming the production, long hours are spent on the set or on location. Actual camera work (filming) may start early in the morning and continue past sunset. Production staff must be knowledgeable and willing to work hard. Members of the production team must be well-trained and use their talents wisely.

COMPUTER/DATA PROCESSING

Computers are electronic devices that receive, change, communicate, and store information. Most modern forms of communication rely on computer systems. They have permitted the development of highly efficient methods of transferring information, Fig. 5-27.

Since the computer industry is so complex, opportunities related to this field are almost limitless. For example, designing and building the actual computer and related equipment (hardware) is a challenging career, Fig. 5-28. Writing internal pro-

grams (instructions for the computer) or user programs (software) is an equally challenging career. These people are called PROGRAMMERS. Most complicated software is developed by computer specialists. Simple programs, however, can easily be written by beginners.

One of the largest uses of computers today is for DATA PROCESSING. In data processing, information (reports, accounts receivable information, client addresses, etc.) is entered into a computer, instructions are entered, and results are received. People who do this specialized work are known as DATA PROCESSORS.

TELECOMMUNICATIONS

Telecommunications involves transmitting information (signals) between distant points. Upon hearing this term, most of us immediately think of telephone networks. However, this industry includes a variety of systems for exchanging messages. The use of satellites to transmit radio signals around the globe is an excellent example, Fig. 5-29. Without communication satellites, we would be unable to receive many television and radio programs.

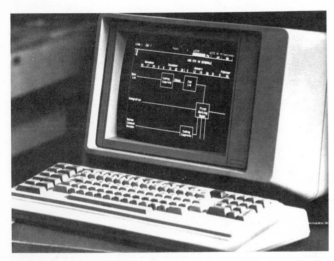

Fig. 5-27. This desktop computer is useful in monitoring production activities in a factory. (Sperry Corp.)

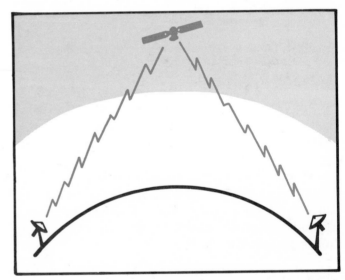

Fig. 5-29. Satellites reflect various communication signals between distant points. That is how we receive TV and radio programs from other continents.

Fig. 5-28. The design and construction of computer equipment is a very complex task. These electrical technicians are checking the machines before shipping. (AccuRay Corp.)

Few people realize how fast modern communication systems are changing. New products and processes are developed almost daily. In fact, many of the systems in place today will be outdated in only a few years. This is true of computer equipment, telephone lines, and other communication devices. The major problem in telecommunications is that new systems are almost out-of-date as soon as they are installed.

Workers in the telecommunications field must have many technical skills. To design radios, telephones, and other equipment demands special training. College degrees in math, science, and engineering are common of telecommunications experts. There are many exciting opportunities for qualified individuals who enter the field, Fig. 5-30.

SERVICE INDUSTRIES

Many firms indirectly provide communication services to consumers. Think about the person who delivers your mail. How does that individual help you communicate with others? Other service-related professions include the local weather forecaster and a taxicab dispatcher. Obviously, the person at the airport control tower has a key communication job.

The largest segment of the service industry involves repair and maintenance work. Many technicians are required to keep communication equipment in order, Fig. 5-31. When complex electronic

Fig. 5-31. Service and repair personnel are well-trained technicians. They must understand very complex machines. (Bell & Howell)

devices need repairing, the bill may exceed several thousand dollars.

The federal government is another service institution. Many of their tasks involve communication technology and systems. Various governmental agencies regulate the flow of information. Departments like the Federal Communications Commission (FCC) oversee the operations of communication companies. Their major task is to set policies and make certain the various companies in the industry follow these policies.

SUMMARY

The communication industry is a large, varied group of companies. These companies all have one thing in common, however. They are organized to create and transmit information, and to be profitable.

In the communication industry, we have identified 11 basic industries. Included are the following: commercial art and design, engineering drafting and design, printing/publishing, photography, advertising, broadcasting, recording, cinematography, computer/data processing, telecommunications, and service.

Each of these groups has a variety of career opportunities. The training or schooling required varies a great deal by position. Many of the positions require a college degree. Other positions require technical training.

KEY WORDS

All the following words have been used in this chapter. Do you know their meanings?

Fig. 5-30. These engineers are assembling a communications satellite for launch. (Rockwell International Corp.)

Advertising agencies, Apprenticeship, Architectural drafters, Communication industries, Companies, Design firms, Drafting, Engineering drawings, Freelance photographer, Industry, Master printers, Publishing houses, Quick printers, Schematic drawings, Special purpose printers, and Technical illustration.

TEST YOUR KNOWLEDGE—Chapter 5

(Please do not write in this text. Place your answers on a separate sheet.)

1. _____ are business enterprises that conduct most economic activity. _____ are a group of related businesses.
2. Commercial design firms:
   a. Specialize in the creative design of commercial items.
   b. Employ layout artists, designers, and printers.
   c. Range in size from one person to hundreds of people.
   d. All of the above.
3. A finely detailed drawing that resembles a photograph is a:
   a. Engineering drawing.
   b. Schematic drawing.
   c. Technical illustration.
   d. None of the above.
4. What two media are closely associated with the broadcasting industry?
5. What is cinematography?

MATCHING QUESTIONS: Match the definition in the left-hand column with the correct term in the right-hand column.

6. __ People who process information through a computer.
7. __ A computer and its related equipment.
8. __ People who write programs.
9. __ User programs.

a. Programmers.
b. Software.
c. Data processors.
d. Hardware.

10. Name the three types of printing companies explained in the text.
11. An _____ specializes in designing the marketing campaigns of other companies.
12. To what industry does the Federal Communications Commission belong?

ACTIVITIES

1. Identify five companies in your community involved in the communications industry. Select one for an in-depth study. How many employees does the company have? What is the amount of their annual sales (dollar amount)? Obtain the name of the person in charge. Write this person a letter, requesting a tour of their facilities.
2. Select an occupation in the communications industry that interests you. It can be one of those mentioned in the chapter, or one that you know about personally. Compile a report on this occupation. What education/training is necessary? What is the future growth of the occupation? (Will it still be around in 15 years?) In what areas of the country are jobs in this occupation located? What is the beginning salary? What is the salary for an experienced professional?
3. Along with your classmates, select one computer or broadcasting company from whom you will buy one share of stock. Follow the value of the stock in the business section of your daily paper. Why does the value change? What factors affect the price of the stock?
4. Make a list of the companies that print your textbooks. Include books from several different subjects/classes. Visit the library to research one or two of these companies. How large are these companies? Do they have only one office? Is all work completed in one location? Summarize your information in a short report.

6 Technical Communication Systems

The information given in this chapter will enable you to:
- *Describe audio, visual, and audio-visual communication systems.*
- *Identify the three technical communication systems we will look at for the remainder of our study.*
- *Cite examples and purposes for these technical communication systems.*

You will recall that communication is the process of exchanging information. The way in which information is exchanged varies a great deal, Fig. 6-1. It may be exchanged simply by talking with a friend. Or the exchange may include two computers "talking" to each other. However this communication takes place, it is a basic activity in daily life.

Our five senses are necessary for communication. They help us recognize and send information. The sense of sight aids us when watching news on television. Many news stories would be difficult to understand without the help of pictures that make up the stories, Fig. 6-2. In addition, our sense of smell can inform us of dangerous situations like a fire. Fires can often be smelled before they are seen. In this case, our sense of smell gives us a very important message. Sometimes the sense of smell communicates pleasant messages, such as "dinner is being prepared."

Fig. 6-2. The control room of a television studio contains many displays and equipment that are necessary for broadcasts.

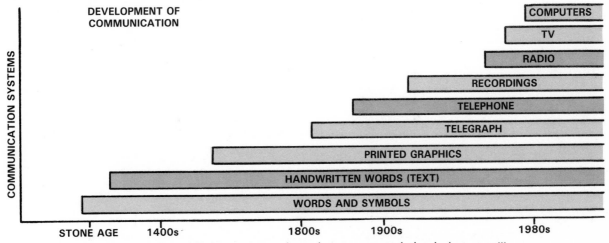

Fig. 6-1. Our ability to communicate has progressed slowly but steadily.

Most (over 90%) of our communication, however, is done with the senses of sight and hearing. These communications are called visual (what can be seen) and audio (what can be heard). Often we use a combination of both sight and hearing. This is referred to as audio-visual communication.

Visual communication is used often in everyday life. Perhaps the most common example of visual communication is the daily paper, Fig. 6-3. We see the print, we read the print, and our minds translate the information being transferred.

A large amount of our messages are audio messages. Whether you listen to a radio or listen to an album, the message is being transferred through the sense of hearing, Fig. 6-4.

Most attempts to communicate information are not simply audio or visual, but both. A high school marching band sends information to be both seen and heard, Fig. 6-5. Using only sight or hearing for this message would result in only a part of the message being received. Other examples of audio, visual, and audio-visual communication are shown in Fig. 6-6.

TECHNICAL COMMUNICATION SYSTEMS

In Chapter 1, communication technology was defined as "the process of transmitting information from a source to a destination using codes and storage systems." This definition includes all types of communication: letter writing, cybernetics, computer programming, etc.

For many of us, the most basic communication is improved with the help of technical devices. A

Fig. 6-3. Newspapers keep millions of people informed every day. They contain a variety of information, from national news to weather.

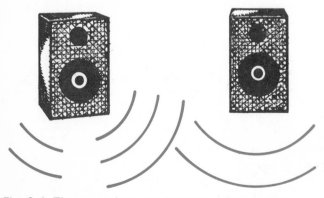
Fig. 6-4. These speakers are the transmitting medium used between a radio and the human ear.

Fig. 6-5. Marching bands must be seen and heard to be enjoyed. (Andy Johnston)

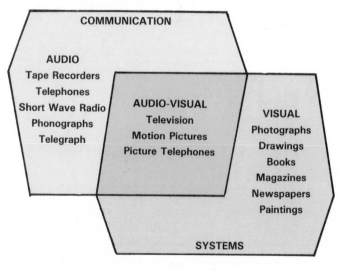
Fig. 6-6. Examples of audio, visual, and audio-visual communication systems.

telephone makes it possible to speak with a friend who lives across town. A label on a candy bar informs us of the nutritional value in the snack.

Now we will look at those communication technologies that use technical devices or systems. We will leave behind topics like public speaking and sign language.

For the remainder of our study, we will take a good look at the areas of graphic and electronic communication. The section on graphic communication will include technical graphics (drawing and drafting) and printed graphics, Fig. 6-7.

TECHNICAL GRAPHICS

How much we depend on sketches and drawings to communicate! Have you ever given directions to others by means of a sketch, Fig. 6-8?

Have you seen graphs or charts used to explain findings? How do simple pictures and directions help in putting a plastic model together?

For the architect and engineer, drawings are the basis for all communication. The design of buildings is explained by technical drawings. Pictorial illustrations communicate ideas to the public. Simple sketches aid in the understanding of ideas and suggestions.

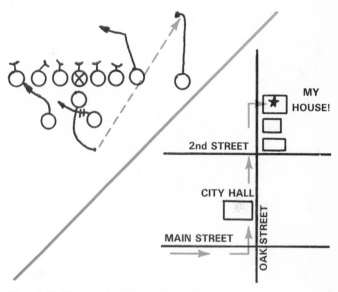

Fig. 6-8. Simple sketches give meaning to many ideas and concepts.

```
SYSTEMS OF
TECHNICAL
COMMUNICATION
 ├── TECHNICAL GRAPHICS
 ├── PRINTED GRAPHICS
 └── ELECTRONIC COMMUNICATION
```

Fig. 6-7. Systems of technical communication.

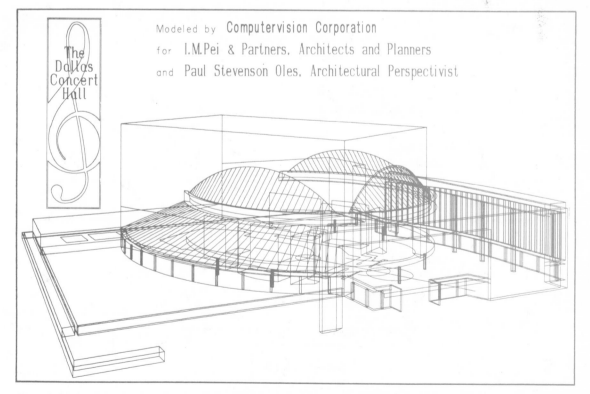

Fig. 6-9. Architectural designers complete drawings of buildings and structures for use during construction. (Computervision Corp.)

The term TECHNICAL GRAPHICS is often used to describe the work of drafters. Technical drawings are prepared with instruments and machines. The most common is a standard engineering drawing or blueprint, Fig. 6-9. Other types of technical graphics are used to show assembly details or pictorial views. At times, engineering and architectural firms use computers to prepare drawings, Fig. 6-10. These graphic systems will be covered fully in Chapter 7.

Fig. 6-10. Modern computers help drafters save hours of time in completing drawings. (Computervision Corp.)

PRINTED GRAPHICS

Communicating with printed images is the most common type of visual communication. Printed graphics appear everywhere, from T-shirts to soda cans. The use of these printed images is commonly called GRAPHIC COMMUNICATION. So, we will now refer to the technology of printed graphics as graphic communication.

Words and pictures printed together transmit information very effectively, Fig. 6-11. The text of this book was printed. Posters, photographs, and large billboards include printed images. What other examples of graphic communication can you identify?

Most graphic messages require careful planning and design. Creating the message involves several layout procedures, Fig. 6-12. The actual printing of the messages is the final step. Printing work is fun and rewarding. But the major purpose of graphic communication is for transmitting messages to others.

The graphic communication field includes activities like photography and photocopying. Photography is an exciting hobby and business. Many people take pictures as a recreational activity, Fig. 6-13. Still others earn incomes by taking wedding and anniversary pictures.

Photocopying also represents a method of

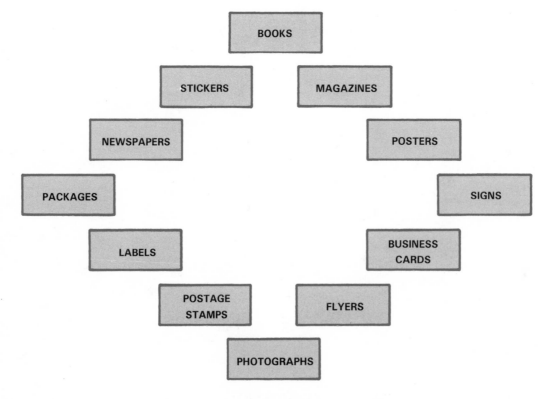

Fig. 6-11. Examples of printed graphics.

Fig. 6-12. Layout artists prepare masters of printed materials. This master will soon be photographed and reproduced by graphic means. (Graphic Products Corp.)

Fig. 6-14. Photocopiers are useful to both schools and businesses. (Xerox Corp.)

signals. This includes systems that rely on electricity to send or receive messages. You probably think of radio and TV signals immediately. But what about radar, computers, and stereo systems? Also consider satellites, telephones, and lasers.

There are many forms of electrical energy used in communication. This is probably because beams (or waves) of light exchange messages quickly and effectively. In fact, telephone calls made and received in the same city are often being sent by light signals. This is called FIBER OPTICS. Waves of acoustic (sound) energy also aid in communication. Your household stereo receiver is a perfect

Fig. 6-13. Photography is a very popular hobby. Many people take pictures of friends or special events.

reproducing materials by graphic means, Fig. 6-14. There are seven major types of printing used today. Several of these techniques may be familiar to you. For example, T-shirts and posters are done by screen process printing. The handouts you receive at school are run on ditto or mimeograph machines. Books are usually printed on lithographic (or offset) presses. These and other graphic communication systems will be explained beginning with Chapter 12.

ELECTRONIC COMMUNICATION

Electronic communication usually represents all transmitting of information by waves, current, or

Fig. 6-15. Both the sound and images used in video games are made electronically.

example. Audio-visual systems are popular too. Video games provide visual images along with audio signals, Fig. 6-15.

Modern computer systems have made some of the largest advances in electronic communication. When first used, computers were large, bulky objects that processed only the simplest information. Desktop models used currently are powerful aids in communicating information, Fig. 6-16. They are used to plan budgets, write letters, or receive the latest information from Wall Street. More complex systems are used by businesses to complete drawings, type letters and labels, and make out paychecks. Electronic communication will be covered beginning with Chapter 18.

SUMMARY

Communication takes place in many ways. In most cases, however, communication is done with the senses of sight and hearing. These senses may be used individually as audio or visual communication. Or they may be used together as audio-visual communication.

Oftentimes audio, visual, or audio-visual communication is strengthened with the help of technical devices. For the remainder of this book we will study three technologies. They are technical graphics, printed graphics (graphic communication), and electronic communication.

KEY WORDS

All the following words have been used in this chapter. Do you know their meanings?

Fig. 6-16. This small, desktop computer is being used to edit and lay out an entire page. (Xerox Corp.)

Acoustic energy, Audio communication, Audio-visual communication, Electronic communication, Fiber optics, Graphic communication, Printed graphics, Technical graphics, and Visual communication.

TEST YOUR KNOWLEDGE—Chapter 6

(Please do not write in the text. Place your answers on a separate sheet.)
1. Over 90% of communication is done with the senses of:
   a. Taste and smell.
   b. Touch and sight.
   c. Sight and hearing.
   d. None of the above.
2. List three examples of audio communication, visual communication, and audio-visual communication.

MATCHING QUESTIONS: Match the definition in the left-hand column with the correct term in the right-hand column.

3. __ Uses information sent by waves, currents, or signals.
4. __ Photography and photocopying are included in this technology.
5. __ Used to describe the work of drafters.

a. Printed graphics.
b. Technical graphics.
c. Electronic communication.

ACTIVITIES

1. Make a list of the electronic communication devices found in your school.
2. Make a poster showing how the picture and sound gets from a television station to a home TV set. Include a brief explanation of the technical system used.
3. Write a short report on the use of computers in your school. Include information on each type of computer and the purpose for which it is used.
4. Arrange for a demonstration of the printing and copying machines used in your school (ditto, photocopier, etc.). Which machines produce the best copies? What is each machine used for?
5. Take two photographs that effectively explain your school's library facility to others.
6. Plan a demonstration of those tools used in technical graphics. What is each tool used for? How is each tool used? Show samples of work done with the various tools.

SECTION 2 TECHNICAL GRAPHICS

7 Introduction To Technical Graphics

The information given in this chapter will enable you to:
○ *Describe the purposes and uses of technical drawings.*
○ *Identify the types of technical drawings used most often in business and industry.*
○ *Discuss the development of technical graphics to its current methods.*

Drawing is a common form of graphic communication. It is one of the easiest and fastest ways to exchange an idea or information. It is often a more effective means of communication than speaking. For example, how would you explain the bones located in the human arm? A verbal description might not completely explain it. A simple drawing, however, can describe what words cannot clearly describe, Fig. 7-1.

Technical graphics also use drawings in order to communicate an idea or message. In addition to the drawing, lines, letters, and symbols are also used, Fig. 7-2.

Details and concepts are also described with the help of technical drawings. Charts and graphs provide a better view of statistics than do a list of numbers. Structural features of buildings are explained on architectural drawings, Fig. 7-3. Products that are manufactured in parts are generally described by working drawings.

Technical drawings are completed by artists and drafters. They produce a variety of drawings, Fig. 7-4. You may have heard them called blueprints

Fig. 7-2. This drafter is completing a technical drawing with the aid of modern computer equipment.
(Computervision Corp.)

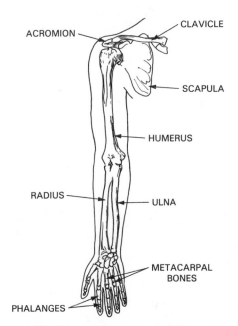

Fig. 7-1. Medical illustrations are useful in explaining parts of the human body.

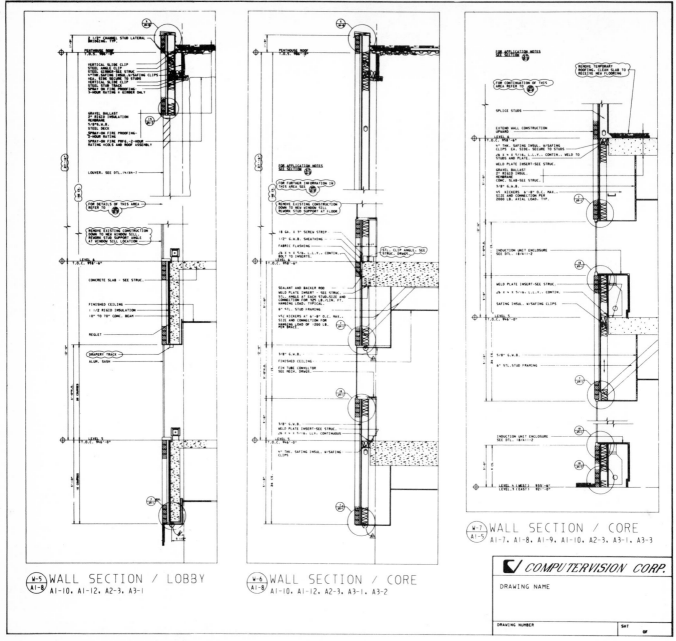

Fig. 7-3. This drawing fully describes construction details for a large building. (Computervision Corp.)

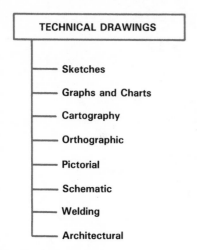

Fig. 7-4. These types of technical drawings are the most commonly used.

or mechanical drawings. Blueprints are actually technical drawings that have been copied on blue paper. Technical drawings are usually developed with mechanical instruments or machines, Fig. 7-5. These tools allow drafters to make finely detailed lines and symbols.

HISTORY OF TECHNICAL GRAPHICS

Long before the use of technical graphics, drawings were being used to communicate. You will recall that the Egyptians used hieroglyphics to communicate ideas, Fig. 7-6.

As tools and machines advanced, people wanted to know how they are built. Some people wanted to learn how to operate the new tools and

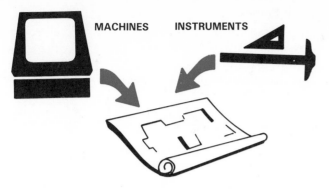

Fig. 7-5. Drawings are often made by using drafting tools or computer equipment.

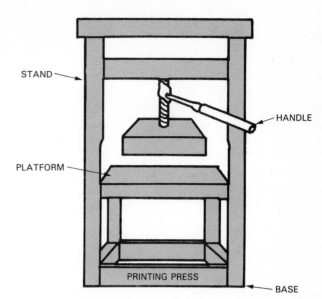

Fig. 7-7. The first printing presses were large pieces of equipment. Sketches identified the major parts of the press.

Fig. 7-6. Hieroglyphics were often made on stone.

machinery. Drawings of these instruments became very important, Fig. 7-7.

The Industrial Revolution brought new demands for drawings. Manufacturers were making machines and tools with many parts. They had to be put together and operated in a certain way. The need for very detailed drawings was the start of technical graphics. Drafters were trained to make the technical drawings that were needed.

Today, the work of drafters is very important to many types of work, Fig. 7-8. Construction

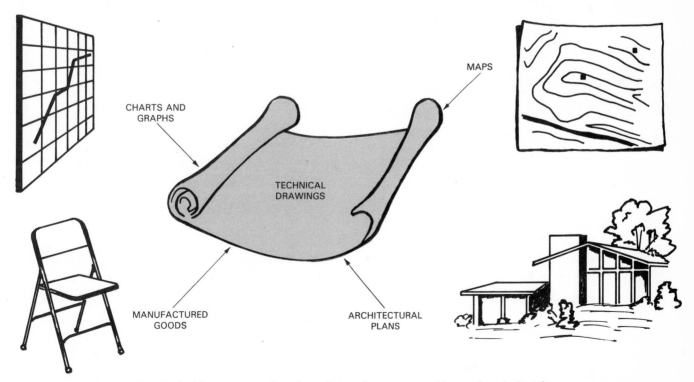

Fig. 7-8. Many types of technical drawings are used in modern industries.

workers need architectural drawings in order to do their jobs. Bankers need charts and graphs to see their statistics more clearly.

The people who make these drawings are highly trained. They must understand the subject they are illustrating. For instance, in order to make a roof detail the drafter must understand architecture. In addition, the drafter must know the best way to present the idea or concept. Will a bar chart be better than a graph? Will a north view be better than a south view? And finally, the drafter must be a creative and skillful artist. Therefore, a drafter's technical education is very important.

TYPES OF TECHNICAL DRAWINGS

SKETCHES

Sketches are often used to communicate ideas or details. Simple sketches are used so often, several types have been developed. The most basic drawing is called a THUMBNAIL SKETCH. This sketch gives simple shapes with only a few lines, Fig. 7-9. As more lines and details are added, the sketch begins to look more like the finished product. These improved thumbnail sketches are known as REFINED SKETCHES. Professional drafters usually complete a refined sketch before continuing their design or artistic work.

GRAPHS AND CHARTS

How often have you seen information or numbers shown in chart form? A large amount of numbers are often organized into a graph or chart. This helps us see and understand important statistics, Fig. 7-10. Graphs are especially useful when explaining business trends, Fig. 7-11.

CARTOGRAPHY

Many areas of land are measured for distance and elevation. Maps of the terrain (land) are then prepared from the measurements taken. This procedure is called CARTOGRAPHY (map making). The people taking the measurements are called SURVEYORS, Fig. 7-12. Their sketches and measurements are given to technical drafters. The

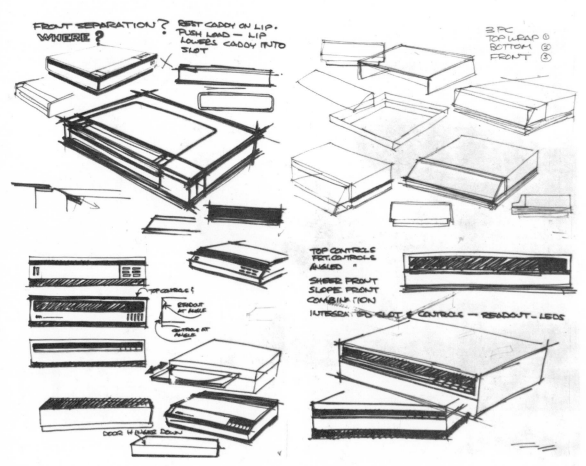

Fig. 7-9. Thumbnail sketches show only the rough outline of an object. Additional ideas are often written next to the sketch. (RCA)

60

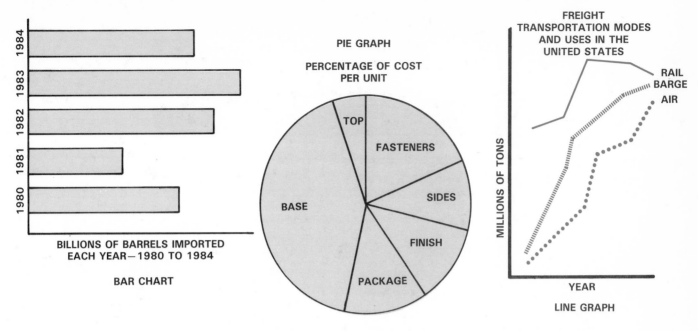

Fig. 7-10. Bar charts, pie graphs, and line graphs are helpful when explaining statistics.

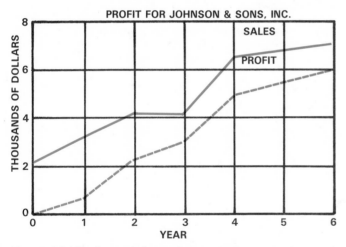

Fig. 7-11. Business information is easier to understand when shown as a chart.

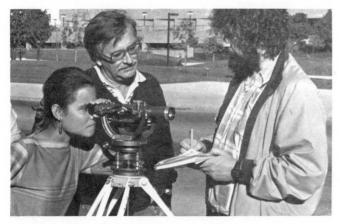

Fig. 7-12. Surveyors measure land areas for distance and elevation. Their measurements are given to drafters, who then prepare maps.

technical drafters then draw a map of the land area, Fig. 7-13.

ORTHOGRAPHIC DRAWING

The major purpose of technical drawings is to describe three-dimensional objects in two dimensions. That means objects are shown on a flat surface, Fig. 7-14. In addition, most items require more than one view to fully describe their features.

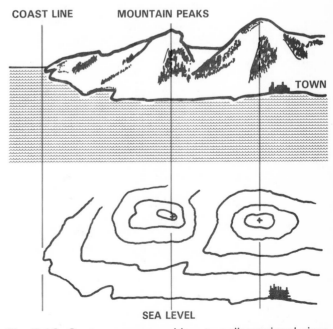

Fig. 7-13. Contour maps provide a two-dimensional view of land areas.

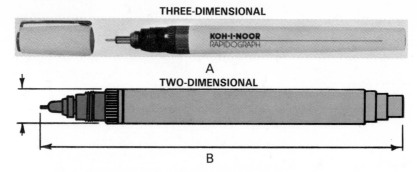

A
TWO-DIMENSIONAL

B

Fig. 7-14. When objects are drawn, only two dimensions are shown—height and width. A—Actual pen is three-dimensional. B—Drawing of the pen is two-dimensional. (Koh-I-Noor Rapidograph, Inc.)

The system of organizing these views is known as ORTHOGRAPHIC DRAWING, Fig. 7-15.

Normally, at least three views are needed to clearly describe any object. This allows the drafter to show the thickness, width, and height of any object. A front, top, and right-side view are the primary views used in orthographic drawing, Fig. 7-16.

Orthographic drawings have several unique features. The most important is the VIEWING PLANE. Every orthographic drawing is developed

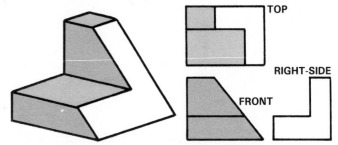

Fig. 7-15. Orthographic drawings generally include a front, top, and right-side view in order to fully describe three-dimensional objects.

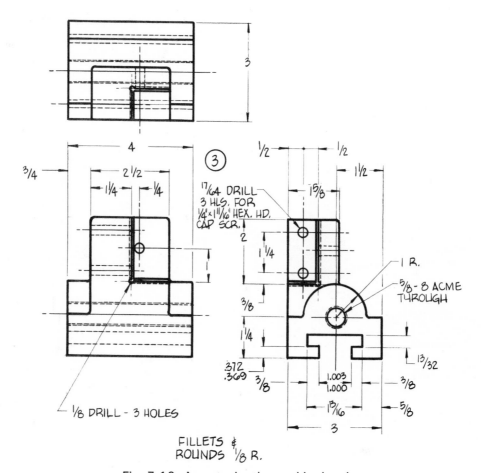

Fig. 7-16. An actual orthographic drawing.

as if a person were looking at the object from an infinite distance. This allows us to visualize how an object looks from a PERPENDICULAR VIEWING PLANE, Fig. 7-17. That means the drawing has the proper height and width but no depth. Another view (top or right-side) is required to show the depth or thickness.

PICTORIAL

Many graphic representations are used to describe objects better than they appear in reality. For instance, it is not always possible to see inside an object. But a pictorial drawing can be used to describe internal features, Fig. 7-18.

Pictorial drawings have many uses in business and industry. Therefore, several types of views were developed, Fig. 7-19. The simplest of these views include ISOMETRIC and OBLIQUE drawings. These show the true length of lines in a single view, Fig. 7-20.

A CUTAWAY VIEW is often used in addition to an orthographic drawing. It helps a drafter to better understand a particular drawing. A cutaway view is shown in Fig. 7-18.

RENDERINGS are usually simple isometric views with shaded areas. The shaded areas make the drawing look more like an actual product. Shading also gives the drawing texture. Methods for shading an object are shown in Fig. 7-21. When done properly, a sample rendering may look quite real. Refer back to Fig. 5-12.

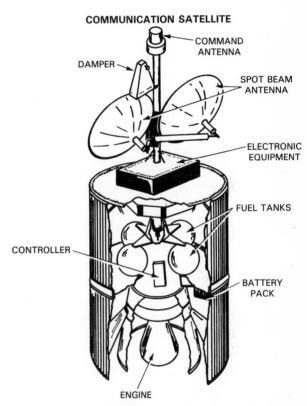

Fig. 7-18. A cutaway view of a communication satellite. This illustrates how useful drawings can be in describing internal parts and details.

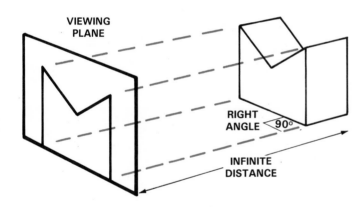

Fig. 7-17. Orthographic viewing planes are perpendicular (at right angles) to each object.

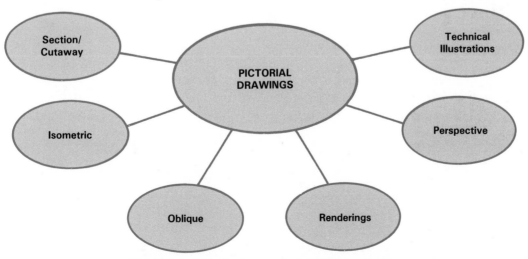

Fig. 7-19. The six types of pictorial drawings.

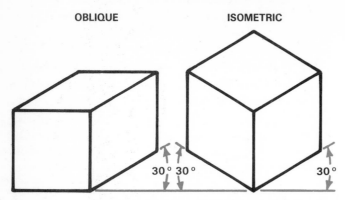

Fig. 7-20. Isometric and oblique drawings. Note the difference in the viewing planes.

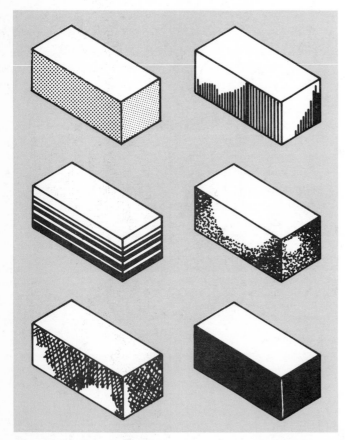

Fig. 7-21. Six methods for shading objects. Shading is especially useful in architectural renderings.

While renderings are used to explain small objects, PERSPECTIVE drawings show large objects or scenes. For instance, an architect may develop a perspective view of a new building, Fig. 7-22.

The final type of pictorial view is called a TECHNICAL ILLUSTRATION. As the name implies, these drawings describe technical devices or systems. A familiar type is an EXPLODED VIEW, Fig. 7-23. A drawing of this type shows how all the parts of an object fit together.

SCHEMATIC DRAWINGS

Technical drawings of electronic devices are called SCHEMATIC DRAWINGS. These drawings show the internal circuitry (wires and electrical components) of various instruments, Fig. 7-24. In this way, electrical systems can be explained more easily. This is especially important in larger electronic systems like radar or stereo equipment. Most electronic devices have thousands of parts. The schematic drawings of these devices may fill many pages.

WELDING DRAWINGS

Welding is a popular production process in the metalworking industry. Welding drawings actually

Fig. 7-22. Perspective drawings provide a realistic view of large structures.

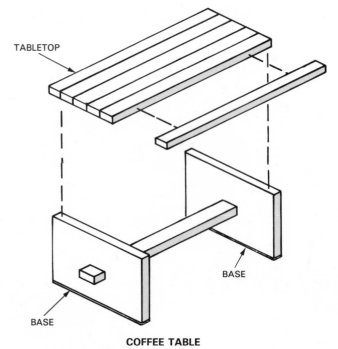

Fig. 7-23. Exploded view illustrations show the way to assemble objects.

serve as instructional plans for where welds are to be placed, Fig. 7-25. This form of drawing demands that the drafter understand welding procedures.

The symbols used in welding drawings are very important. These symbols have been developed by the American Welding Society (AWS). Drafters use care when using various welding symbols on drawings, Fig. 7-26. Use of an improper symbol may be dangerous.

ARCHITECTURAL DRAWINGS

The planning of buildings and homes is much like the planning of manufactured goods. Drafters must develop detailed drawings of each part of the structure. Houses, stores, skyscrapers, and other buildings are all designed on paper before construction begins.

Architectural drawings come in various forms. In fact, an entire set of drawings is required for a typical building, Fig. 7-27. Technical drawings of each system are completed to describe different details.

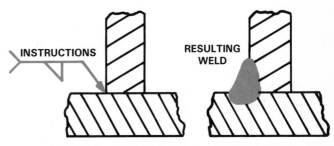

Fig. 7-25. Welding drawings give instructions about where and how metal parts are to be welded.

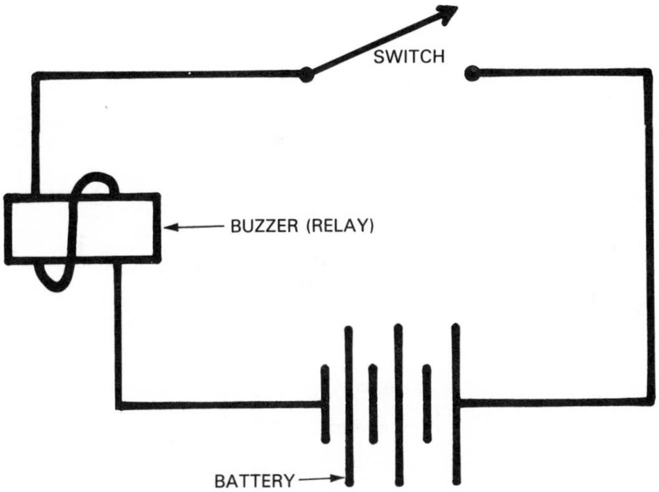

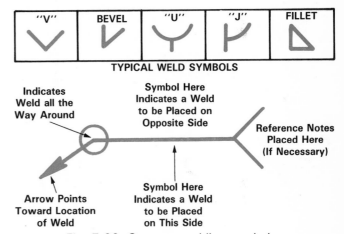

Fig. 7-26. Common welding symbols.

Fig. 7-24. This schematic view of a doorbell explains how the circuitry is put together.

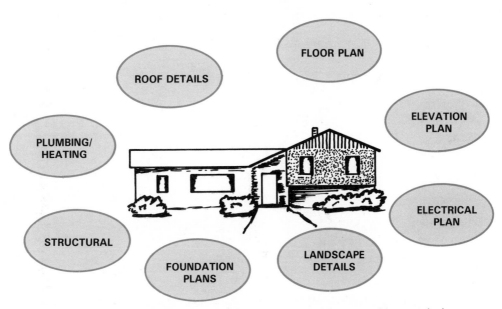

Fig. 7-27. Every detail of a building is described in an architectural plan.

The most familiar type of architectural drawing is called a floor plan, Fig. 7-28. Other drawings cover foundation, structural, roof, plumbing, and ventilation plans. All plans are equally as important. Complete and accurate drawings are necessary for proper construction, Fig. 7-29.

SUMMARY

The purpose of technical graphics is to help the communication process in business and industry. In many cases, these drawings present ideas and objects more clearly than a verbal explanation.

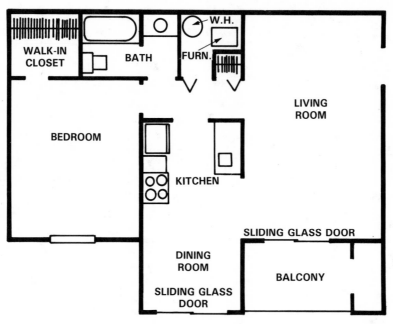

Fig. 7-28. Floor plans show the layout of rooms in homes or offices.

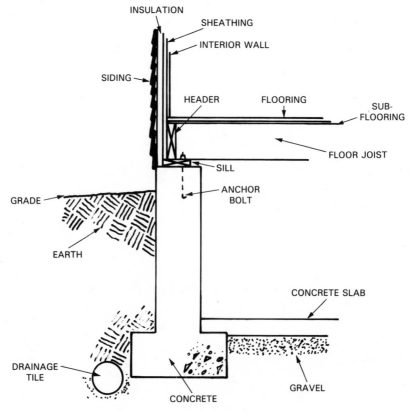

Fig. 7-29. Foundation details of buildings are often shown as section views.

The need for technical graphics came from our growing technology. Since the Industrial Revolution, technical graphics have become very important. Large, complicated machines require detailed drawings for production and use.

The most common of technical drawings are sketches, graphs and charts, cartography, orthographic drawings, pictorial drawings, schematic drawings, welding drawings, and architectural drawings.

KEY WORDS

All the following words have been used in this chapter. Do you know their meanings?

Architectural drawing, Blueprints, Cartography, Chart, Cutaway view, Exploded view, Graph, Isometric drawing, Mechanical drawing, Oblique drawing, Orthographic drawing, Perspective drawing, Refined sketch, Renderings, Schematic drawing, Surveyors, Technical illustration, Thumbnail sketch, Viewing plane, and Welding drawing.

TEST YOUR KNOWLEDGE — Chapter 7

(Please do not write in this text. Place your answers on a separate sheet.)
1. Technical graphics use drawings to communicate:
   a. A message.
   b. Details.
   c. Concepts.
   d. All of the above.
   e. None of the above.
2. _____ are actually technical drawings that have been copied onto blue paper.
3. The ability to draw is the only skill technical drafters need. True or False?
4. The people who take measurements for maps are called:
   a. Drafters.
   b. Architects.
   c. Surveyors.
   d. None of the above.

MATCHING QUESTIONS: Match the definition in the left-hand column with the correct term in the right-hand column.
5. __ Pictorial view shaded to add texture.
6. __ Drawing of electronic circuitry.
7. __ Map making.
8. __ Drawing of houses or commercial buildings.
9. __ Sketches made with a few lines.
10. __ Drawing with multiple views.

a. Thumbnail.
b. Cartography.
c. Rendering.
d. Architectural.
e. Schematic.
f. Orthographic.

11. When a _____ viewing plane is used, drawings have the proper _____ and _____ but no depth.
12. Name three types of pictorial drawings.

ACTIVITIES

1. Secure the architectural drawings used for your school building. Identify the plumbing details, foundation plans, roof details, etc.
2. Practice shading sketches by using several of the methods presented in this chapter.
3. Collect a sample of several of the drawings discussed in this chapter. What is each drawing used for? If possible, obtain a photograph of the finished product.
4. Develop a thumbnail sketch of a bicycle. Then refine the sketch. Finally, make an artist's rendering of the bike.
5. Invite a surveyor to visit your class. Discuss the various maps that were made from measurements taken by the surveyor. What do the different shadings, lines, and symbols describe? Ask the surveyor about his/her job. What training or education is required?
6. Gather information from your local newspaper that would work well as a chart or graph. Use this information to design either a bar chart, pie graph, or line graph. Explain your chart or graph to the class.

8 | Technical Graphics Procedures

The information given in this chapter will enable you to:

○ *Become aware of freehand sketching techniques used in technical communication.*

○ *Understand the uses of sketching in design and in planning activities.*

○ *Identify the tools and machines commonly used in technical graphics.*

○ *Communicate with others by means of simple sketches and diagrams.*

As you learned in Chapter 7, there are many different types of technical drawings. Each type of drawing is used for a particular situation. For example, a pictorial drawing may not always be the best way to communicate an idea. Perhaps a schematic drawing would work best.

While technical drawings are different, the way they are made is basically the same. We will examine three of these procedures in this chapter.

They are: freehand sketching, drawing with instruments, and computer-aided drafting, Fig. 8-1.

FREEHAND SKETCHING

Technical communication starts with individuals attempting to solve problems or develop designs. While they "think through" different possibilities, many solutions come to mind. We call this practice VISUALIZATION, Fig. 8-2. As plans or designs are developed, we like to communicate our thoughts. This leads to several options for exchanging the information.

A simple method of exchanging ideas is speaking. However, a better method is by sketching the ideas, Fig. 8-3. Simple pictures may fully explain one's thoughts. If more details are needed, a formal drawing can be done.

Freehand sketching is often the only communication required to transmit ideas. In this case, the exchange of information is complete. How-

A

B

C

Fig. 8-1. Technical drawings are completed by three different means. A—Freehand. B—With instruments. C—By machine.

ever, the sketch cannot just be thrown together. It is important that the sketch be developed correctly. The placement of notes on the sketch is also important. See Fig. 8-4.

Fig. 8-2. We try to tackle problems by visualizing possible solutions in our minds.

Fig. 8-3. It is easier to describe objects with pictures than with words.

TOOLS AND MATERIALS

Freehand sketching requires very few tools or supplies. That makes this form of communication useful in many instances. Sketches may be completed anywhere a pencil and paper are located. Chalk and a blackboard also work well for freehand sketching, Fig. 8-5.

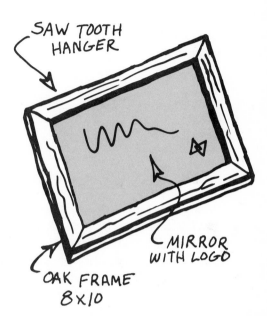

Fig. 8-4. Even the simplest sketch is more descriptive when notes are added.

Fig. 8-5. Sketching is easily done on either paper or chalkboards.

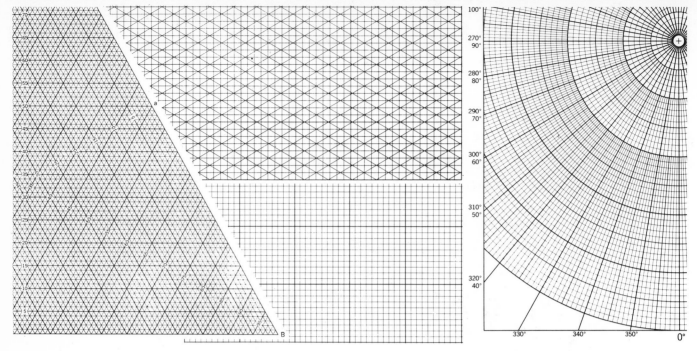

Fig. 8-6. Graph paper is a valuable aid in producing freehand sketches.

GRAPH PAPER is helpful in making simple sketches, Fig. 8-6. The grid (lines) guides the drafter in keeping lines straight, parallel, or of equal length. Various types of graph paper are available in stores. One of the most popular graph papers with technical drafters is called ISO-METRIC PAPER, Fig. 8-7.

Sketching is done with a variety of pencils. Standard lead pencils work well for thumbnail sketches. When cleaner lines are required, a MECHANICAL LEAD HOLDER is helpful. Mechanical holders are more convenient because the lead rarely needs sharpening. The lead in the pencil is a standard width and hardness, Fig. 8-8.

Fig. 8-8. Mechanical lead holders are used to draw sharp, clean lines.

FREEHAND TECHNIQUES

The best drawings are the result of careful and accurate sketching techniques. Placing your paper on a hard, flat, smooth surface is important. In addition, the writing surface should be at a comfortable height. All sketching work should be done in a relaxed position, Fig. 8-9.

It is wise to sketch very light lines at first, Fig. 8-10. Major lines are darkened as the drawing develops. With practice, simple drawings can be completed with ease.

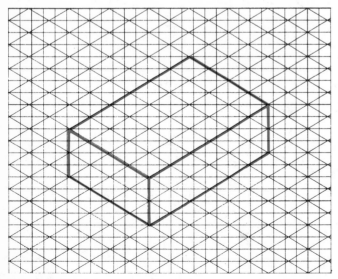

Fig. 8-7. One of the many ways isometric (grid) graph paper is used.

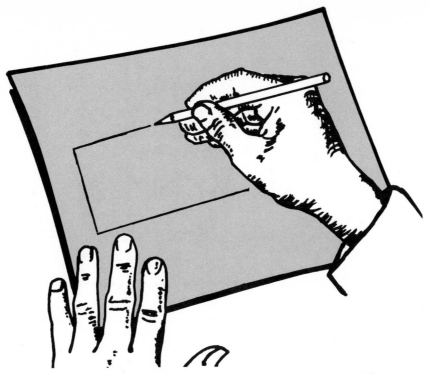

Fig. 8-9. Sketching should be done in a comfortable position.

HAND LETTERING

Notes provide important information on technical sketches and drawings, Fig. 8-11. Examples include labeling colors, materials, or important dimensions. As sketches develop, the notes become more important. Thumbnail sketches may have only rough outlines and no labeling. Refined sketches are fully explained with the use of notes or other information, Fig. 8-12.

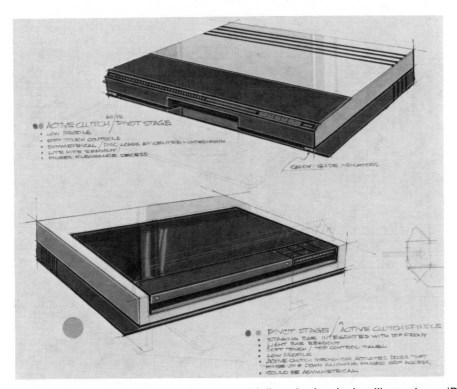

Fig. 8-10. Very light lines often serve as guidelines in developing illustrations. (RCA)

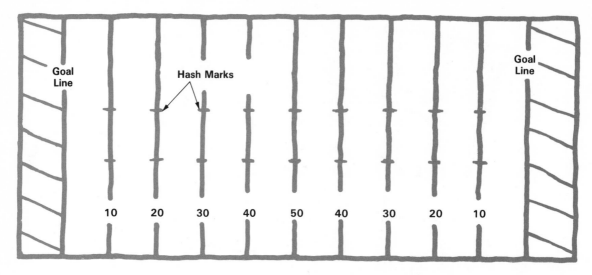

Fig. 8-11. Notes and dimensions are required to fully explain most sketches.

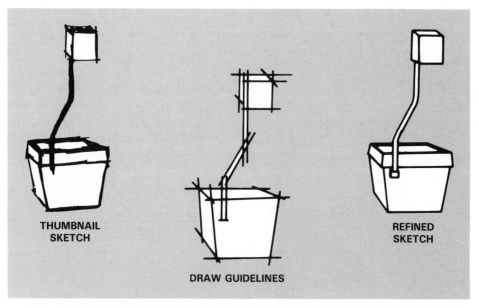

Fig. 8-12. Refined sketches are usually developed from thumbnail views.

DRAWING WITH INSTRUMENTS

Standard technical drawings are made with a variety of instruments and machines, Fig. 8-13. These drafting instruments allow individuals to create finely detailed drawings.

DRAFTING INSTRUMENTS

Among the most commonly used instruments are the T-SQUARE and the DRAFTING BOARD, Fig. 8-14. Other common instruments include triangles and engineering scales (ruler). A list of important tools appear in Fig. 8-15.

T-squares and triangles are used to draw straight, parallel lines. When circles or curved lines are required, different tools are available. FRENCH (or irregular) CURVES are one tool used for drawing these lines. A regular compass is useful in drawing circles. TEMPLATES are also used in creating any curved or circular images. See Fig. 8-16.

Remember that technical drawings must be prepared as neatly as possible. Therefore, typical equipment includes a pencil sharpener, erasers, and a dusting brush. Sharp pencils permit the drafter to draw and label each drawing in a professional manner, Fig. 8-17. Erasers and dusting

Fig. 8-13. This drafter is using mechanical instruments to complete a certificate. (Letterguide, Inc.)

Fig. 8-14. A typical drafting board and instruments.

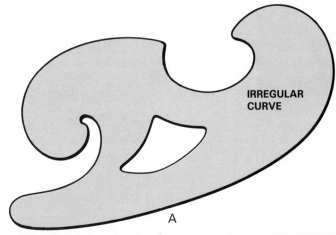

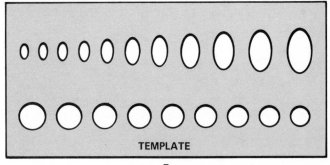

Fig. 8-16. Two instruments that are useful for drawing circles or curved lines. A—French (irregular) curve. B—Templates.

Fig. 8-17. Electric sharpeners keep pencil leads sharp. (Koh-I-Noor Rapidograph, Inc.)

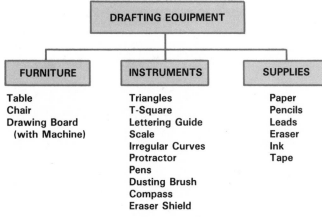

Fig. 8-15. Mechanical drafting equipment.

DRAFTING EQUIPMENT

FURNITURE	INSTRUMENTS	SUPPLIES
Table	Triangles	Paper
Chair	T-Square	Pencils
Drawing Board	Lettering Guide	Leads
(with Machine)	Scale	Eraser
	Irregular Curves	Ink
	Protractor	Tape
	Pens	
	Dusting Brush	
	Compass	
	Eraser Shield	

brushes help maintain a clean working surface.

Lettering guides are helpful in labeling technical drawings. When using a lead pencil, guidelines may be constructed with a T-square or Ames lettering guide. Special templates are useful when inking a drawing, Fig. 8-18. Lettering guides help lessen the problem of smeared or smudged notes.

A number of lettering styles are acceptable for technical drawings. Most lettering guides are flexible and allow drafters to produce several styles. Examples of lettering styles are illustrated in Fig. 8-19.

DRAFTING INSTRUMENT TECHNIQUES

Using drafting instruments usually requires some instruction. While these instruments are not difficult to use, they do require practice.

As with freehand sketching, it is important that the drawing surface be smooth and flat. The edges of the drafting board must be straight. This keeps the T-square aligned. A drafting machine can be used in place of a T-square, Fig. 8-20. Again, this will require a bit more practice than using a T-square.

COMPUTER-AIDED DRAFTING

Many industries need drawings completed quickly and efficiently. Modern computers help drafters create technical presentations with ease, Fig. 8-21.

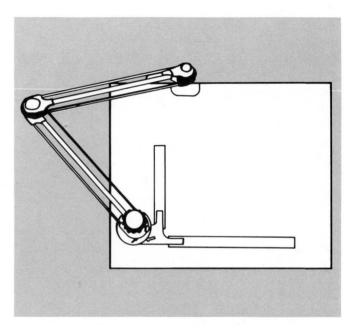

Fig. 8-20. A drafting machine.

Fig. 8-18. Neat label drawings are made with the help of lettering guides. (Letterguide, Inc.)

Hand Lettered Mechanical Style

ABCDE abcdef 0123456789

Hand Lettered Architectural Style

ABCDE abcde 123456789

Mechanical Lettering

ABCDE abcdefgh 0123456789

Italic Mechanical Style

ABCDEF abcdefg 0123456789

Fig. 8-19. Styles of technical lettering.

Fig. 8-21. Computers are used in many industries to aid designers and drafters. (General Motors Corp.)

The use of computers to develop drawings is known as COMPUTER GRAPHICS. A more familiar term is computer-aided drafting or CAD. (The letters CAD may also stand for computer-assisted drafting and computer-aided design.)

There are many good reasons to use computers for drawing. They perform simple functions (like drawing) with great accuracy and speed. Information can be stored internally or on disks. This means standard drawings only need to be done once. After the drawing is done it is stored in the computer for future use.

Computer-aided drafting is quite different from instrument drawing. T-squares and other tools are not required. No lines are drawn by hand or instrument either. Instead, commands are entered on a keyboard, or MENU PAD, Fig. 8-22. Computers operate on the instructions entered by keyboard or menu. These detailed instructions are called PROGRAMS. Computers must be programmed to produce drawings on the screen. Even the simplest drawings require a complex set of instructions for the computer, Fig. 8-23.

EQUIPMENT

Technical drawing can be done on most computer systems. Standard systems consist of a keyboard, microprocessor (the computer), and a screen. A printer or plotter is often included. These allow the programmer to transfer the drawings to

Fig. 8-22. Computer commands may be entered through a keyboard or with a menu pad.

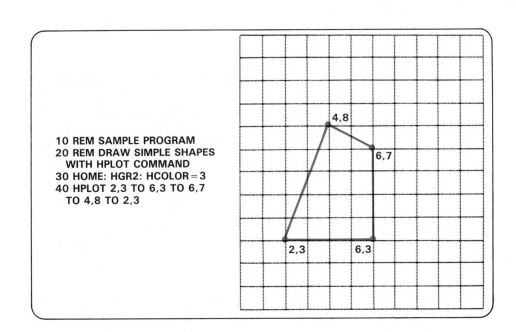

```
10 REM SAMPLE PROGRAM
20 REM DRAW SIMPLE SHAPES
   WITH HPLOT COMMAND
30 HOME: HGR2: HCOLOR=3
40 HPLOT 2,3 TO 6,3 TO 6,7
   TO 4,8 TO 2,3
```

Fig. 8-23. A sample computer program. This program instructs the computer to draw a geometric figure.

paper. All the equipment just mentioned is called HARDWARE, Fig. 8-24.

The most important requirement for CAD work is the program, or SOFTWARE. Many types of software are available commercially. Programs are usually written in a variety of languages for the different systems found in industry. The most popular computer languages are BASIC, Fortran, and Pascal.

Computer programs may be stored in a variety of ways. The most common storage mediums include disk and magnetic tape, Fig. 8-25. A disk drive is required to load programs from computer disks. Tape-recorded programs are loaded from cassettes or cartridges.

CAD TECHNIQUES

Computer-aided drafting hardware and software vary by system. Large industrial equipment allows drafters to complete very complex drawings. The computer systems found in typical schools are much smaller and easier to use. The techniques for creating drawings are different for each computer. We will now work through a typical example.

An operating program must be loaded into a computer's memory to operate any software. OPERATING PROGRAMS allow the drafter to make lines, circles, and labels on the screen. The commands entered by the user are known as a program or APPLICATION SOFTWARE. This program varies for each drawing developed. For example, a program to draw a house is different than one to draw a tree, Fig. 8-26.

Modern software offers many advantages for technical drafters. One bonus is color. Complex drawings may be developed with each part or

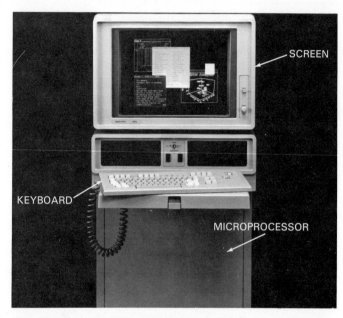

Fig. 8-24. Computer equipment is known as hardware. Study the type of hardware in this system. (Tektronix, Inc.)

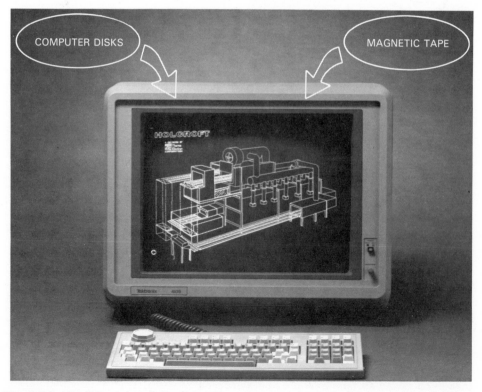

Fig. 8-25. Technical drawings may be stored on computer disks or magnetic tapes.

system drawn in a different color. Another convenience is the ability to draw standard symbols easily. Most operating software has routine symbols and figures already programmed. By pressing a single key, circles or labels are drawn automatically, Fig. 8-27.

The final result of all this programming is a drawing. It must change from an image on a screen to a drawing on paper. Computer PRINTERS or PLOTTERS allow programmers to create a hard copy. HARD COPY is the drawing produced by a computer. Industrial printers can develop complex drawings on paper in minutes, Fig. 8-28.

SUMMARY

There are three ways to create technical drawings. Each procedure requires different levels of time and training, and different types of tools.

The first method for making technical drawings is freehand sketching. Development and notes are important in order to complete a good sketch. Very few tools are required for freehand sketching. In addition, training time is short.

The second method for producing technical drawings is drawing with instruments. Drafting boards and T-squares are the two most important

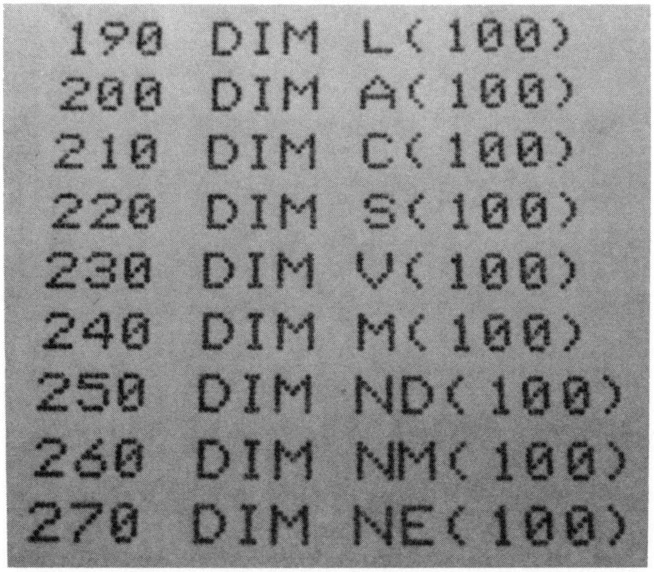

```
190 DIM L(100)
200 DIM A(100)
210 DIM C(100)
220 DIM S(100)
230 DIM V(100)
240 DIM M(100)
250 DIM ND(100)
260 DIM NM(100)
270 DIM NE(100)
```

Fig. 8-26. Application software serves as instructions for a drawing.

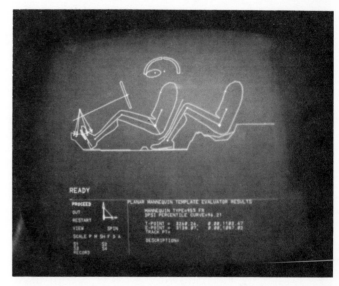

Fig. 8-27. This computer is used in the auto industry. It has internal programs to draw standard shapes, like people and car seats. (General Motors Corp.)

Fig. 8-28. A hard copy of the technical drawing shown on the screen is being printed on the left. (Bausch & Lomb)

tools for this method. A fair amount of instruction and practice are necessary in order to use this method.

The last method for making technical drawings is computer-aided drafting, or CAD. CAD has the longest training time of the three methods. Advanced schooling and technical training are necessary to be a designer or programmer in this field. In addition, it requires the use of many instruments, tools, and machines.

KEY WORDS

All the following words have been used in this chapter. Do you know their meanings?

Applications software, Computer-aided drafting (CAD), Computer graphics, Drafting board, French curve, Graph paper, Hard copy, Hardware, Isometric paper, Keyboard, Lettering guide, Mechanical lead holder, Menu pad, Plotters, Printers, Program, Templates, Triangles, T-square, and Visualization.

TEST YOUR KNOWLEDGE—Chapter 8

(Please do not write in this text. Place your answers on a separate sheet.)

1. _____ is the practice of thinking through a problem.
2. Sketching may be done almost anywhere with little equipment or supplies. True or False?
3. Which of the following is NOT a tool used in sketching?
   a. Grid paper.
   b. Mechanical lead holder.
   c. Pencil.
   d. Plotter.
4. CAD stands for:
   a. Computer-aided drafting.
   b. Computer-aided design.
   c. Computer-assisted drafting.
   d. All of the above.
   e. None of the above.
5. _____ _____ prevent smudged or smeared letters from occurring.

MATCHING QUESTIONS: Match the definition in the left-hand column with the correct term in the right-hand column.

6. __ Equipment used in CAD.
7. __ Used to enter commands into a computer.
8. __ Instructions for the computer to follow.
9. __ The drawing produced by a computer.
10. __ Used to create the hard copy.

a. Program.
b. Menu pad.
c. Plotter.
d. Hardware.
e. Hard copy.

ACTIVITIES

1. Assemble a collection of mechanical drafting instruments. Display them on a table. On index cards, write a description of each tool and what the tool is used for. Display these cards next to the proper instrument.
2. Draw a picture of a cardboard box freehand and with drafting instruments. How do they differ? What might each particular drawing be used for? If possible, have the cardboard box drawn on a computer. How does this differ from the other two drawings?
3. Obtain samples of the various lettering styles. Using several of the styles, print your name and address.
4. Arrange to tour a local business that uses CAD. What does the business use it for? How many people are needed to operate this system? What is the quality of the graphics made by CAD?
5. Research the latest CAD equipment. Recent technical magazines are a good resource. Present a short report on your findings.
6. Draw a drafting board and T-square on isometric grid graph paper. Observe all the proper procedures for making a drawing with instruments.

9 Technical Sketching and Drawing

The information given in this chapter will enable you to:
- ○ *Discover the purposes and uses of sketching in technical communication.*
- ○ *Understand the skills necessary to prepare sketches and drawings.*
- ○ *Produce technical sketches and drawings using the various sketching techniques.*

In most instances, sketching is considered a common form of freehand drawing. Simple lines and figures are drawn without the aid of instruments, Fig. 9-1. Freehand sketches are often used as a guide for laying out technical drawings. Basically, simple symbols and shapes can be drawn either by hand or machine. The processes outlined here apply to many forms of graphic communication.

There are two basic forms of sketching done today. The first is TWO-DIMENSIONAL. Two-dimensional sketches only show the height and width of an object. Most engineering drawings are two-dimensional. The other basic sketching form is a PICTORIAL SKETCH. A pictorial sketch includes the depth of the item, as well as the height and width. Illustrations of this type are THREE-DIMENSIONAL. Examples of both forms are shown in Fig. 9-2.

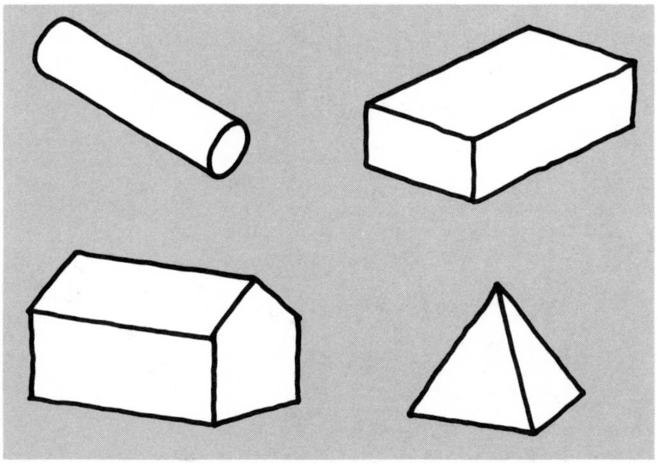

PURPOSES AND USES

Sketches are used often as a part of daily work. How often do your teachers draw simple figures on the chalkboard, Fig. 9-3? The pictures of designs communicate an idea or information. In much the same way, industrial personnel explain details to others with sketches. They use simple drawings to explain the floor plan of a house or to show a plan for solving a problem.

Technical sketching and drawing is a vital part of many industries. Designers and drafters develop a variety of illustrations for every technical drawing they produce. Sketches are made of early ideas to evaluate possible designs more easily. As improvements are discovered, the sketches are refined. Careful review of these rough drawings allows many options to be considered. In addition,

Fig. 9-1. It would be very difficult to explain any of these figures without the use of sketches.

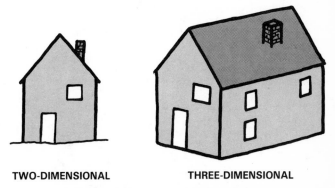

TWO-DIMENSIONAL   THREE-DIMENSIONAL

Fig. 9-2. A two-dimensional view shows only the height and width of the house. When a third dimension is added, the depth of the house can also be seen.

Technical Sketching and Drawing 79

others may be asked to look at the preliminary sketches. Their suggestions might aid in improving or completing the design work.

Finally, drawings of the final plans must be organized and prepared. A sketch of how the illustrations will be arranged is helpful, Fig. 9-4. Light guidelines also aid in constructing the lines on the drafting paper.

SKETCHING TECHNIQUES

Sketching is the simplest form of graphic communication. Whether you are sketching on a piece of paper, a chalkboard, or a computer screen, only two pieces of equipment are required, Fig. 9-5. The skills needed to produce simple sketches are basic. They are outlined in the following section.

LINES

We generally think of a line as being a fairly basic object. There is only one kind of line and it is drawn with a ruler. This is not always the case with sketching. There are many types of lines used in graphic work, Fig. 9-6. Each one has a purpose and must be used properly to avoid confusion.

GUIDELINES are used to give a sketch its basic shape. These lines are drawn very lightly with a pencil.

OBJECT LINES are used to show the visible

Fig. 9-3. Even math problems and hopscotch squares are types of sketches.

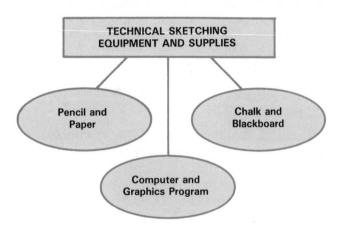

Fig. 9-5. The equipment needed to make a technical sketch is very basic.

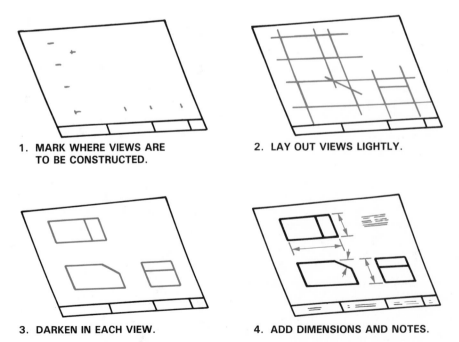

1. MARK WHERE VIEWS ARE TO BE CONSTRUCTED.

2. LAY OUT VIEWS LIGHTLY.

3. DARKEN IN EACH VIEW.

4. ADD DIMENSIONS AND NOTES.

Fig. 9-4. The stages of developing a mechanical drawing.

80

edges of an object. These lines are heavy and thick.

HIDDEN LINES are used to show edges or parts of an object that are not visible. They are drawn using a series of dashes, with spaces between the dashes.

DIMENSION LINES are used to show the dimensions of an object. They usually have arrowheads at the ends of the lines. Dimension lines are very fine lines.

CENTER LINES are used to find the centers of objects. They are made up of a series of long and short dashes with spaces in between.

SKETCHING LINES

Generally, most of the lines used when sketching are drawn either horizontally or vertically. Sketching either of these is fairly simple, Fig. 9-7.

To sketch a HORIZONTAL LINE, mark off two points. The distance between these points should be equal to the length of the line you wish to draw, Fig. 9-7A. In addition, the points should be located parallel to the top or bottom of your sheet of paper. Next, connect these points with a guideline, Fig. 9-7B. Move your pencil back and forth lightly until the guideline is complete. Finally, sketch an object line over your guidelines, Fig. 9-7C. Move the pencil in ONE motion, from left to right.

Sketching a VERTICAL LINE is very similar to sketching a horizontal line. Again, mark off two points, making sure the distance between the two points is equal to the length of the line to be drawn. The two points, however, should be made parallel to the right or left edge of your sheet of paper. Follow the same procedure used for horizontal lines when making the guideline and the object line.

Three-dimensional sketches also use horizontal and vertical lines. However, many times these lines are drawn at angles that are not completely parallel with the edge of the sheet of paper. For example, depth lines do not run parallel to the edge of a sheet, Fig. 9-8. These lines are called INCLINED LINES.

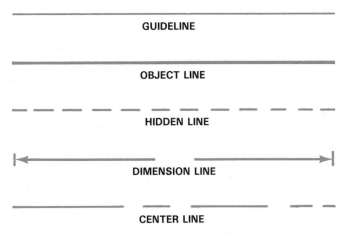

GUIDELINE

OBJECT LINE

HIDDEN LINE

DIMENSION LINE

CENTER LINE

Fig. 9-6. Five types of lines used most often in technical sketching.

DESIRED LINE

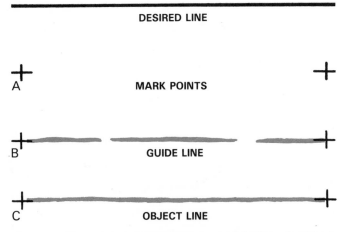

A — MARK POINTS

B — GUIDE LINE

C — OBJECT LINE

Fig. 9-7. Method for constructing a straight line. A—Mark off end points equal to length of desired line. B—Sketch guidelines. C—Sketch object line.

INCLINED LINES

Fig. 9-8. Three-dimensional sketches have many inclined lines. Can you find any others in this sketch?

Sketching these lines is, once again, similar to sketching horizontal or vertical lines. When marking off the two points, place them at the desired angle. Connect them with a guideline. Then sketch over the guideline with an object line. When the line inclines (goes up) to the right, sketch up. When the line inclines to the left, sketch down, Fig. 9-9.

CIRCLES

Now that you have learned how to sketch horizontal, vertical, and inclined lines, you are ready to sketch circles and other curved surfaces. This is because horizontal, vertical, and inclined lines are used to sketch a circle.

When sketching a circle, the first step is to sketch a series of horizontal, vertical, and inclined lines. All these lines should cross one another at a common midpoint. Next, on each line mark off a unit that is equal to the radius of the desired circle. When each line is marked, connect each point to the next point with guidelines. Finish by darkening with an object line, Fig. 9-10.

Many sketches contain details that appear circular, Fig. 9-11. In most cases it is better to draw the circular shapes first. These shapes can be used as a gauge for sketching the rest of the object.

DETAILED SHAPES

Geometric shapes are important in technical graphics. Technical drafters must be able to create these shapes quickly and easily, Fig. 9-12. That is because these basic figures are used for making more detailed shapes.

Many figures are formed like standard shapes such as squares, triangles, or circles, Fig. 9-13. This is important to remember when sketching detailed objects. If the basic shape can be determined, it can be used as a guide. For example, a five-point star is made using a group of triangles, Fig. 9-14. After lightly sketching the group of triangles, the object lines are darkened. You have now drawn a star. Try drawing the same star without using triangles.

Fig. 9-9. Sketching inclined lines requires practice. Try this practice exercise on a sheet of paper.

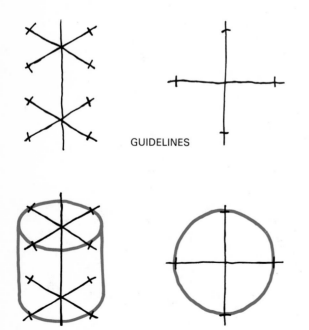

GUIDELINES

Fig. 9-10. Sketching curved lines is easier with the use of guidelines.

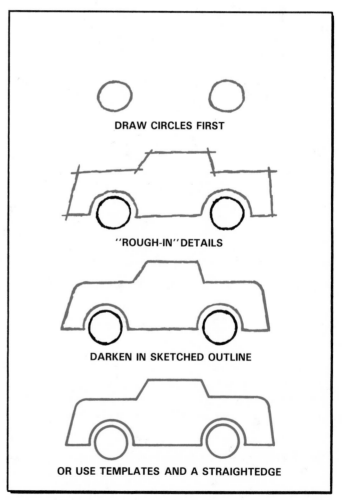

DRAW CIRCLES FIRST

"ROUGH-IN" DETAILS

DARKEN IN SKETCHED OUTLINE

OR USE TEMPLATES AND A STRAIGHTEDGE

Fig. 9-11. Making a sketch of a car. Sketching the tires first makes sketching the remainder of the car easier.

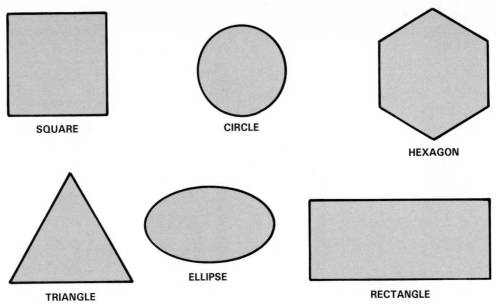

Fig. 9-12. These common geometric shapes are often used to form detailed shapes.

SQUARE

CIRCLE

HEXAGON

TRIANGLE

ELLIPSE

RECTANGLE

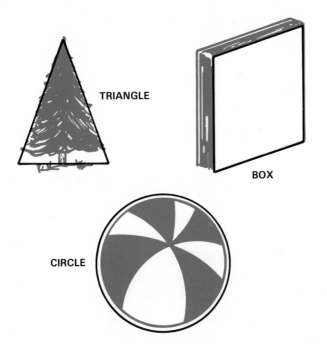

TRIANGLE

BOX

CIRCLE

Fig. 9-13. Most objects can be reduced to a few familiar shapes.

Architects often plan buildings by sketching how the structure will look upon completion. They use standard shapes to lay out the overall design, Fig. 9-15. You may have noticed that the overall shape of a building is the total of several standard shapes.

PROPORTION IN SKETCHING

Drawings are often enlarged (made bigger) or reduced (made smaller) to fit the needs of the person using them. For instance, a particular detail

might be easier to read if drawn on a larger scale, Fig. 9-16. The size of a sketch may be changed by using a GRID SYSTEM.

Changing the scale of a sketch requires an understanding of the term PROPORTION. Objects are often compared by physical size. A drawing that is exactly twice the size of another drawing is proportionally "double-sized," Fig. 9-17.

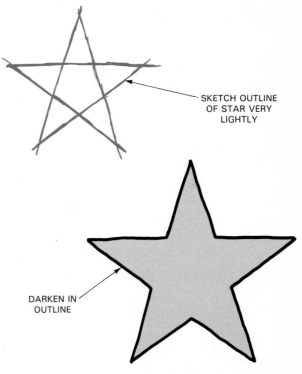

SKETCH OUTLINE OF STAR VERY LIGHTLY

DARKEN IN OUTLINE

Fig. 9-14. Simple shapes are often sketched first, before the actual object is drawn in with darker lines.

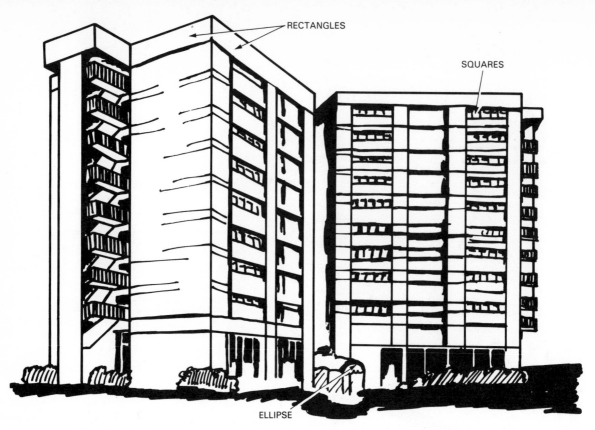

Fig. 9-15. The design of a building is the result of combining simple shapes.

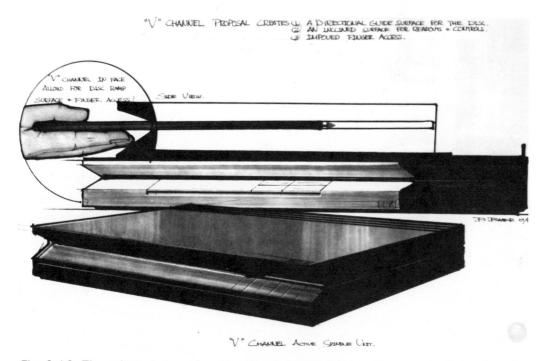

"V" CHANNEL PROPOSAL CREATES ① A DIRECTIONAL GUIDE SURFACE FOR THE DISC.
② AN INCLINED SURFACE FOR READOUTS + CONTROLS.
③ IMPROVED FINGER ACCESS.

"V" CHANNEL IN FACE
ALLOWS FOR DISC RAMP
SURFACE + FINGER ACCESS!   SIDE VIEW.

"V" CHANNEL ACTIVE SPINDLE UNIT.

Fig. 9-16. The enlarged view of a video disc being loaded into the recorder helps describe the technique. (RCA)

Sketches like these can be drawn very easily. The key is in developing a grid that is enlarged or reduced to the proper size or scale.

The first step in changing the size of a drawing is to block off the original drawing into a grid.

These squares should be drawn lightly and be of the SAME SIZE. Lines are spaced every 1/4 inch to 2 inches, depending on the size of the original. Next, a second grid is drawn to the desired scale. For example, if the original drawing has squares

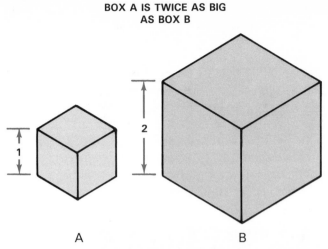

BOX A IS TWICE AS BIG AS BOX B

A B

Fig. 9-17. Proportion is the relationship of sizes among several objects.

1/2 inch in size and the enlargement is to be twice as large, the second grid will have squares 1 inch in size. Study Fig. 9-18. Carefully note where the object lines cross the grid markings in the original drawing. These points are transferred onto the larger grid, in the same place. Then, using the points as a guide, the shape is sketched into the grid. The outline is darkened with object lines to complete the sketch.

Just as drawings can be enlarged, they can be reduced. The same procedure is followed. The difference is that the second grid is made smaller than the original grid.

COMPUTER GENERATED SKETCHING

In the last chapter, you discovered how computers help with sketching and drafting. Sketching may be done on computers as well as paper. In fact, it may be easier "drawing" on a screen. Modern computer systems permit the development of very detailed three-dimensional views, Fig. 9-19.

Computers and software can vary a great deal by manufacturer. Different companies often design their computers and software differently. That means the techniques (commands) used to develop lines will differ. However, one procedure remains fairly consistent. A program will have to be entered into the computer. This set of commands will "tell" the machine how to produce the image on the screen, Fig. 9-20.

Most graphic programs reduce a drawing to a series of points. You may have followed this procedure when sketching the solution to a math problem on graph paper, Fig. 9-21. When a line is created, the end points are determined and entered in the computer. Then the machine

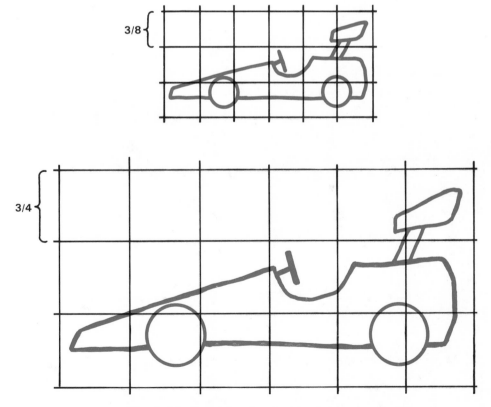

Fig. 9-18. Example of a double-sized enlargement.

calculates all points between the end points of the line. Therefore, to produce a drawing on a computer, you must understand how to describe the shape with numbers. This is an important process in technical design work, Fig. 9-22.

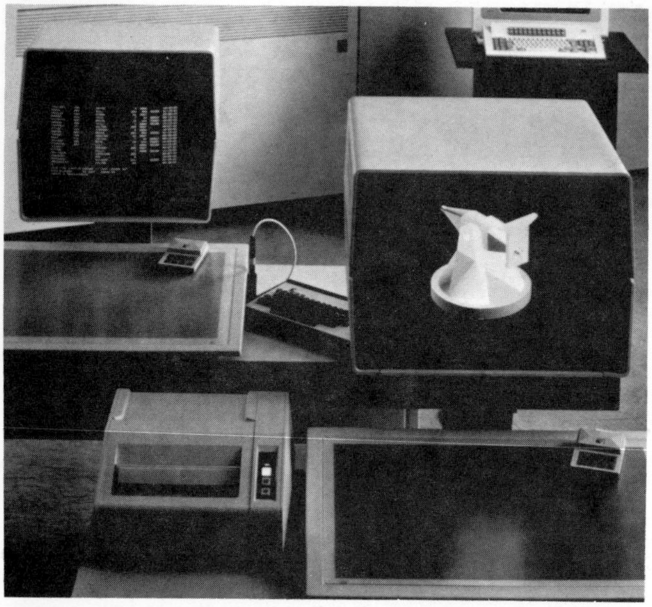

The number of points along a line varies by the computer graphics program used. Systems with HIGH-RESOLUTION capabilities produce more detailed lines than a LOW-RESOLUTION system. RESOLUTION is determined by the number of "dots" that can be drawn per inch, Fig. 9-23. More points along a line results in a smoother line or curve. Microcomputers found in most schools have to be "programmed" to enter a high-

Fig. 9-19. A three-dimensional view is shown on the computer monitor at the right. Standard text (on the screen at the left) is an example of a two-dimensional image. (Computervision Corp.)

Fig. 9-20. This computer has two screens. A program may be entered on one monitor, while the resulting drawing is shown at the right. (Ball State University)

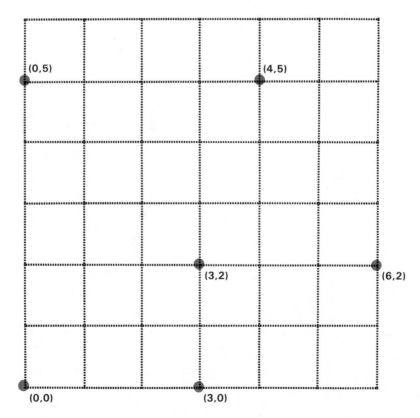

Fig. 9-21. Plotting points on graph paper is similar to a computer plotting points on a computer screen.

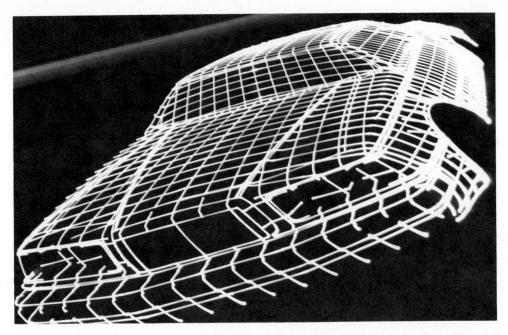

Fig. 9-22. This view of a new automobile has been reduced to simple points and lines. A laser is used to sketch the view for easy reference. (Ford Motor Co.)

resolution mode.

The newest computer systems allow drafters to produce drawings very efficiently. For example, a circle can be drawn by entering a center point and radius. The computer "draws" the image instantly on the screen. Triangles, rectangles, and other standard shapes are developed just as easily. A starting point is entered along with a height and width dimension. The computer system creates the shapes either on a screen or print-out.

Computer graphics programs have other convenient features for developing drawings, too. One advantage is in labeling a sketch or drawing. Dimensions and notes can be added easily by keyboard. This replaces the need to hand-letter all text and numbers, Fig. 9-24. Another convenient feature is the ability to change the size (or scale) of various details. The sizes of a view can be

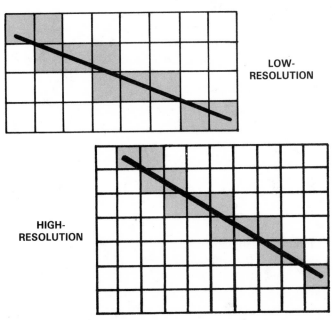

LOW-RESOLUTION

HIGH-RESOLUTION

Fig. 9-23. Resolution refers to the number of dots per inch. High-resolution lines are created by placing more dots along a line.

Fig. 9-24. Computers can label drawings faster and neater than can be done by hand. This system allows the user to point to the letters and symbols that should be drawn on the screen.

Technical Sketching and Drawing 87

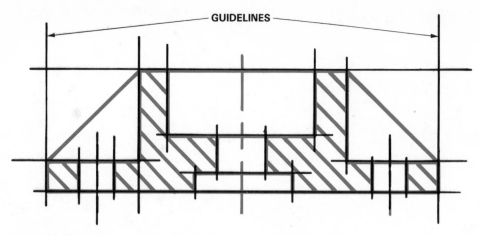

GUIDELINES

Fig. 9-25. A complete two-dimensional technical drawing. This is a cutaway view.

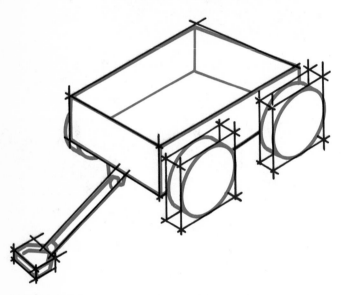

Fig. 9-26. What common shapes do you see in this pictorial sketch?

enlarged or reduced. Another convenience is the ability to "move" a drawing to a different location or page. This command also helps in moving a view to the center of the screen or paper.

DEVELOPING VIEWS

All of the sketching techniques just discussed go together towards developing a view. It is important to do the techniques correctly. This will insure that the views are also developed correctly. Constructing lines, shapes, and figures in the proper way is important. When done properly, the result is an accurate technical drawing.

The procedure for developing two-dimensional views often includes sketching lines. Generally, an entire drawing is sketched first. This includes sketching both an outline and any internal features. Then the major object lines are darkened

as desired, Fig. 9-25. This sketch will serve as a guide for developing the view.

Pictorial views almost always start as sketches. Artists and drafters usually begin with a detailed sketch. When developing views, they look for familiar shapes, Fig. 9-26. Light lines are then drawn to show the outline or shape. As more detail is included, the drawing looks more realistic.

SUMMARY

Everyone uses sketches at one time or another. These sketches help us communicate with others. They visually describe objects and scenes. Whether on paper or on a computer screen, sketching is an important procedure.

Sketches and drawings are produced in stages. Early ideas are generally sketched in a very rough state. Little equipment is required for these illustrations. Detailed shapes can be sketched by following light guidelines. Sketches are complete when all major lines are darkened as needed.

Sketches developed by computer graphics systems are much the same. Lines and shapes must be organized on the screen. In most cases, a program will be used that "tells" the computer how to draw a view. A print-out of the drawing can be produced on paper.

KEY WORDS

All the following words have been used in this chapter. Do you know their meanings?

Center lines, Dimension lines, Enlargement, Grid system, Guidelines, Hidden lines, High-resolution, Horizontal lines, Low-resolution, Object lines, Pictorial sketch, Proportion, Reduction, Three-dimensional, Two-dimensional, and Vertical line.

TEST YOUR KNOWLEDGE—Chapter 9

(Please do not write in this text. Place your answers on a separate sheet.)

1. Two-dimensional sketches show only the _____ and _____ of an object. Three-dimensional sketches also show the _____.

MATCHING QUESTIONS: Match the definition in the left-hand column with the correct term in the right-hand column.

2. __ Usually has arrowheads at the end of the line.
3. __ Shows the visible edges of an object.
4. __ Used to give a sketch its basic shape.
5. __ Show parts of object that are not visible.

   a. Guideline.
   b. Hidden line.
   c. Dimension line.
   d. Object line.

6. What is the only difference in procedure for sketching horizontal lines and vertical lines?
7. When sketching, it is best to draw circular shapes first. True or False?
8. Define proportion.

ACTIVITIES

1. Develop sketches of the following: a pyramid, a soda can, a textbook, and an overhead projector. Briefly explain to the class the techniques you used for each sketch.
2. Make two sketches of a sailboat; one should be exactly twice the size of the other.
3. Invite a newspaper illustrator to visit your class. Ask for a demonstration of the different techniques discussed in this chapter. What kind of education is necessary to become an illustrator?
4. Collect a variety of freehand sketches. Separate them into two groups: two-dimensional and three-dimensional. Explain, in your opinion, the purpose of each sketch.
5. Sketch the Olympic Games symbol (5 rings) in several different sizes. Show how you reduced or enlarged each sketch.

10 Orthographic Drawing

The information given in this chapter will enable you to:
- *Become aware of the uses of orthographic sketching and drawing.*
- *Identify typical systems of sketching and coding used in orthographic drawings.*
- *Understand techniques for developing orthographic projections.*

You already understand the importance of communication in modern society. This is especially true in the industrial sector. Manufacturers and builders must depend on accurate communication. Suppose a designer produced an incomplete blueprint of your family car or an architect failed to include key details in your school's plans. These unsuccessful attempts to communicate vital information could be very dangerous.

Modern industries rely on various forms of visual communication to exchange information and ideas. This may include typed letters, photographs, computer images, or technical drawings. Presently, sketching and drawing represents a major part of technical communication. Technical graphics has become both useful and necessary in industry, Fig. 10-1.

In earlier chapters, you learned about different types of technical drawings. Basic equipment and supplies were also explained. This section will ex-

Fig. 10-1. These architects work late into the night to complete plans for a new building in their community. (Teledyne Post)

plore methods of preparing orthographic drawings. You will also discover the purposes behind orthographic drawings.

All technical drawings use a universal (worldwide) system for communicating ideas. That way the symbols and methods of layout are easily recognized worldwide. We often refer to this as "following a CONVENTION," Fig. 10-2. The major systems established for technical graphics are the American National Standards Institute (ANSI) and the Aerospace-Automotive Drawing Standards (SAE).

PURPOSES AND USES

As products and structures become more complex, designers face increased pressure. Drafters and designers are responsible for developing the plans for all details and parts, Fig. 10-3. Much of their work goes into the technical drawings used during production.

Generally, design drawings illustrate the size and shape of objects. Related information about materials and function may also be included. These drawings served as instructions for how the item is to be produced, Fig. 10-4. Manufactured products often require many pages of drawings to adequately describe all parts. Blueprints for large buildings may fill several books.

Technical drawings are usually superior to photographs. Much more information can be given in a drawing. Sketches and drawings can describe shapes more accurately. Internal details may be illustrated in sectional views. Different features might be drawn in a larger scale to show fine details.

Plans and other design information are used by

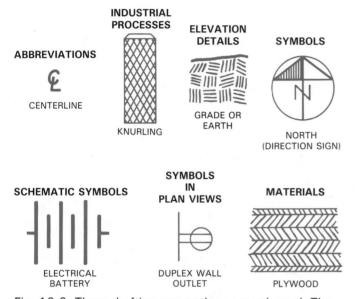

Fig. 10-2. These drafting conventions are universal. They mean the same thing to all drafters and architects.

Fig. 10-3. Production of television sets requires thousands of drawings to specify shapes, sizes, and other details. (Zenith Electronics Corp.)

many people in a typical manufacturing firm. The technical drawings show shop personnel what is to be produced. After review of the drawings, facilities and machines can be designed. Materials can be ordered based on the plans and accompanying bill (list) of materials. During production, assembly drawings guide workers in putting parts together properly.

Architectural drawings serve basically the same purpose. Materials and supplies are ordered after reading the architect's plans. The blueprints inform construction workers how to build the building or structure. Additional drawings describe details such as plumbing and heating ducts. Drawings for commercial buildings often include plans for landscaping or other exterior work. A set of complete architectural plans fills many pages, Fig. 10-5.

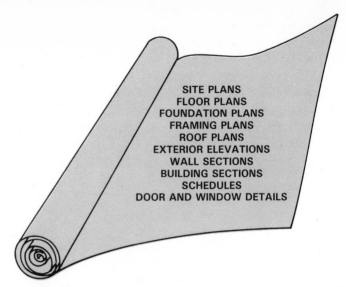

SITE PLANS
FLOOR PLANS
FOUNDATION PLANS
FRAMING PLANS
ROOF PLANS
EXTERIOR ELEVATIONS
WALL SECTIONS
BUILDING SECTIONS
SCHEDULES
DOOR AND WINDOW DETAILS

Fig. 10-5. Ordinarily, architectural plans for a building or structure will fill many pages.

Fig. 10-4. This production manager checks a design drawing that shows how an automobile is to be welded together. (Chrysler Corp.)

ORTHOGRAPHIC PROJECTION

The primary type of drawing used in industry is called an orthographic projection. The term MULTIVIEW DRAWING is also used to describe these drawings. ORTHOGRAPHIC PROJECTION refers to the procedure of projecting images from an object, Fig. 10-6. In order to project this view, a POINT OF REFERENCE is established at an infinite distance. Generally, the point will be perpendicular (at a right angle) to the object. This point provides the ORTHOGRAPHIC VIEW. A drawing is then made from the view shown on the reference plane.

In most cases, several views are necessary to describe any object. That is where the term multiview becomes important. Several drawings (or a multiple number) are constructed to ade-

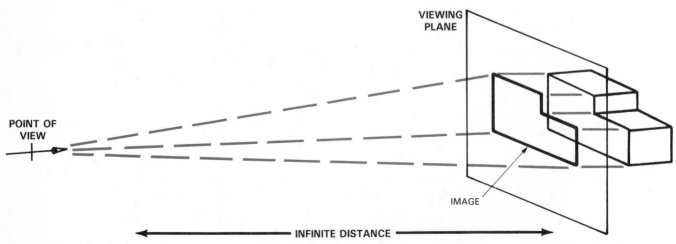

VIEWING PLANE

POINT OF VIEW

IMAGE

INFINITE DISTANCE

Fig. 10-6. Lines projected from an object (intersecting a reference plane) form an orthographic image.

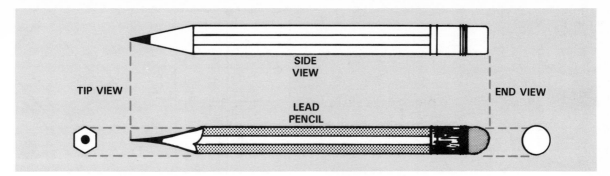

Fig. 10-7. Orthographic drawings contain several views of the same object, as seen from different reference points.

quately illustrate an item, Fig. 10-7. A standard set of orthographic drawings includes at least three views. Front, top, and right-side views are the most popular. A typical multiview is shown in Fig. 10-8.

ORGANIZATION OF VIEWS

Orthographic drawings have many advantages. None is more important than the ability to completely describe an object on a single, flat surface. Several views can be shown together, Fig. 10-9. Even hidden details may be noted.

As mentioned earlier, three views are projected in an orthographic drawing. A front view is constructed off the primary surface. This view generally provides the best view of the object, Fig. 10-10. To adequately describe the entire shape and size, other views are required. Both a top and right-side view are often used to complete the projection. Since these views are sometimes identical, one may be unnecessary, Fig. 10-11.

The placement of each view in orthographic projection is important. All drawings are located in

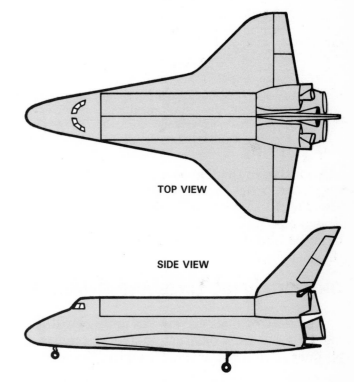

Fig. 10-9. Several views of an object can be shown in an orthographic drawing.

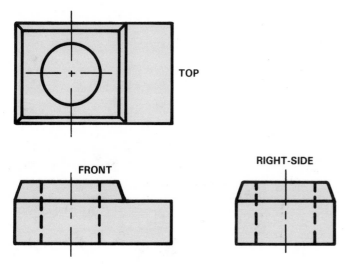

Fig. 10-8. A typical orthographic, or multiview, drawing.

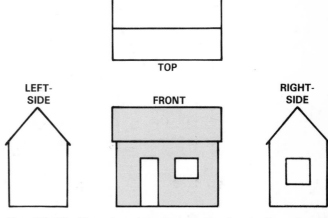

Fig. 10-10. The view selected as the front view should show the most detail.

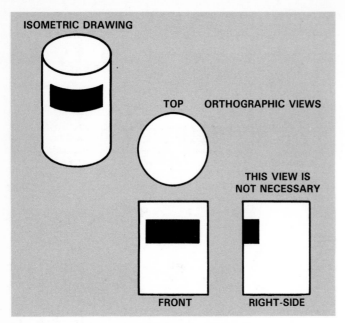

Fig. 10-11. Many objects appear the same from different angles. In this case, there is no need for more than two drawings.

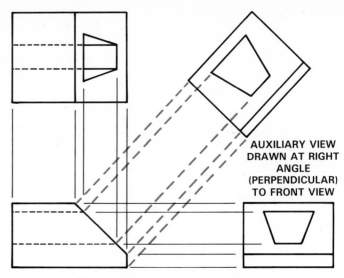

Fig. 10-12. Orthographic views are developed in direct relationship to the front view.

relationship to the front (or major) surface. This means a top view is drawn directly above the object, Fig. 10-12. Left- and right-side views are positioned beside the front view. Auxiliary views that describe odd details are drawn perpendicular to those surfaces.

ORTHOGRAPHIC LINES AND SYMBOLS

A variety of lines are used in multiview drawings, Fig. 10-13. Construction and placement of all lines is vital for accuracy. Object lines on all objects are drawn with bold (heavy) lines. Hidden features are shown with broken (dashed) lines. Center and dimension lines, and symbols are generally drawn with thin lines. Notes and dimensions are also drawn with thin lines.

Dimension lines and notes should be made carefully. Sizes and distances are labeled with dimension lines, Fig. 10-14. Leaders extend to, but never touch, the dimensioned object. Arrow heads show the exact distance being identified.

Internal features of many objects are important in overall function or design. A view may be drawn as if the object has been cut in two, Fig. 10-15. Section lines (A--A) show where the cut is made. The adjoining view is cross-hatched to illustrate solid areas. If several parts are identified, cross-

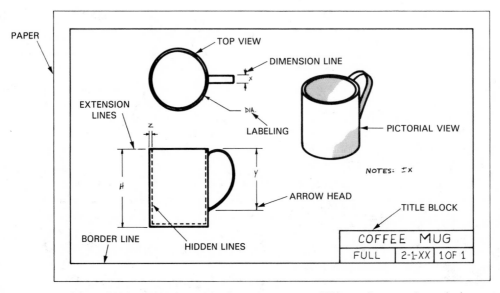

Fig. 10-13. Multiview drawings use many different lines and symbols.

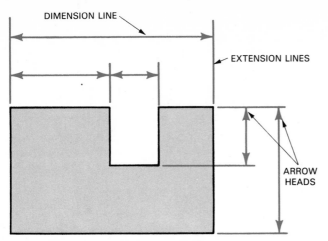

Fig. 10-14. Parts of a dimension line.

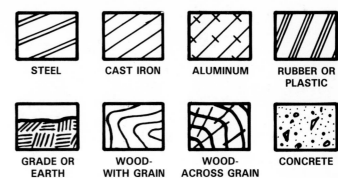

STEEL   CAST IRON   ALUMINUM   RUBBER OR PLASTIC

GRADE OR EARTH   WOOD-WITH GRAIN   WOOD-ACROSS GRAIN   CONCRETE

Fig. 10-16. Many materials are represented by a unique symbol. This is a sample of technical symbols used for sectional views.

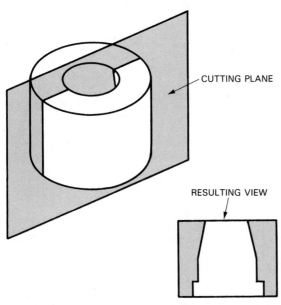

CUTTING PLANE

RESULTING VIEW

Fig. 10-15. Sectional (cutaway) views show the internal features of objects. Drawings like this allow you to see important shapes in the center of an item.

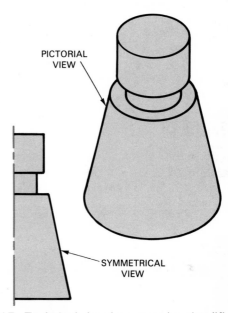

PICTORIAL VIEW

SYMMETRICAL VIEW

Fig. 10-17. Technical drawings may be simplified when their shape is symmetric. A centerline shows the plane of symmetry in this object.

hatching is done at different angles for each part. In some cases, the cross-hatching represents the material used, Fig. 10-16.

ORTHOGRAPHIC CONVENTIONS

Many conventions are useful in developing orthographic views. One method of representation deals with SYMMETRY. As you know, many objects appear identical on both sides or both ends. That means one side of the drawing will resemble the other exactly (symmetry), Fig. 10-17. Rather than drawing both sides, a centerline is developed to show the symmetry. This saves time and space.

Another technique that reduces the total area needed involves BREAK LINES. Little is gained

in drawing full lengths of pipes, I-beams, or other long objects. Instead, lines showing a break in the total distance are used, Fig. 10-18. Very long distances can be shown without leaving out important features. Large details can be shown in a much smaller space. This will save room on the drafting paper.

Many objects are often difficult to draw. A perfect example is the threads on the end of a bolt. Rather than draw every thread, the bolt may be described with symbols and notes. This saves time without adding confusion, Fig. 10-19. The sizes of threads are easily marked with simple notes.

NOTES AND DIMENSIONS

Dimensions and notes are often as important as the actual views in many technical drawings.

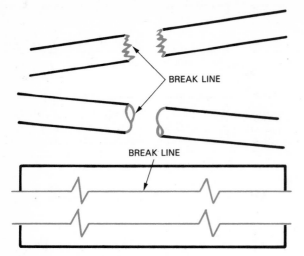

Fig. 10-18. Break lines are used in views of long objects or distances. This saves room in the drawing, while maintaining all the necessary details.

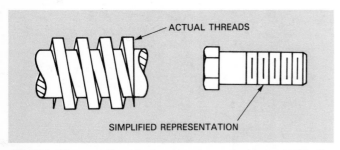

Fig. 10-19. Drawing the threads of a typical bolt would take hours to do correctly. This method of representing the thread pattern is much quicker, easier, and understandable.

Sizes, distances, and features must be identified to complete an accurate drawing. Labeling the drawing is a slow but necessary task, Fig. 10-20. There are many rules for using dimensions and notes on orthographic drawings.

Fig. 10-20. Adding notes and dimensions to drawings is usually a slow, demanding task. Lettering guides are helpful aids. (Letterguide, Inc.)

Generally, labeling is placed on the most descriptive views or features, Fig. 10-21. Major sizes or distances are listed on the front view. It is recommended that holes be marked in a top view as a visible feature, rather than as hidden lines. Depth may be labled in either the side or top views. One word of caution: never label the exact size of a line unless it is shown as full length.

Dimension lines and leaders must be constructed neatly around a drawing. Dimension lines should

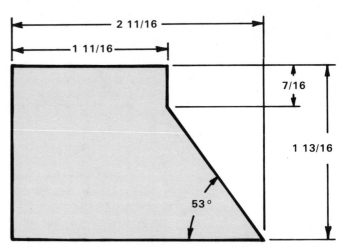

Fig. 10-21. Dimensions and notes should be carefully placed on each view. Place labeling on the view that best describes the feature being marked.

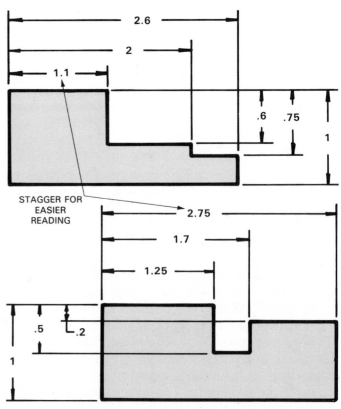

Fig. 10-22. Dimensions are often crowded around a view. These numbers were staggered to allow for easier reading.

not be overlapped, unless necessary. Staggering the dimensions (numbers) will "loosen up" a particular view, Fig. 10-22. In certain cases, a few dimensions might be moved to another view. This helps reduce the clutter around a view that requires many dimensions.

There are different methods for placing dimensions on technical drawings, Fig. 10-23. Standard figures, decimals, and fractions generally provide enough information to communicate sizes and distances. When sizes are allowed to vary slightly, several dimensions are listed. The range between these sizes is called the TOLERANCE. The tolerance range listed on drawings gives the maximum or minimum sizes permitted.

Finally, lettering must be added to the drawing. Light guidelines are useful in lettering technical views. Templates or lettering guides also help when lettering a page, Fig. 10-24. If a computer graphics program is being used, all lettering is done by machine. In fact, the computer may add dimension lines and values, too. Modern programs allow drawings to be labeled easily, Fig. 10-25.

DEVELOPING ORTHOGRAPHIC DRAWINGS

Technical or engineering drawings are often developed on large sheets of drafting paper. Many objects can be shown on the same page. However, a drafter must carefully select the best views to use on the drawing. Organization of the drawing is important. By properly laying out the views, dozens of orthographic views may be placed on one sheet.

The procedure for organizing orthographic views is quite simple. You should already know the exact sizes of the objects to be drawn. A

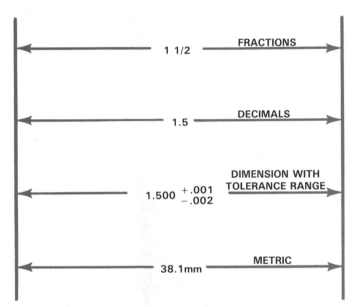

Fig. 10-23. Methods of dimensioning on a technical drawing.

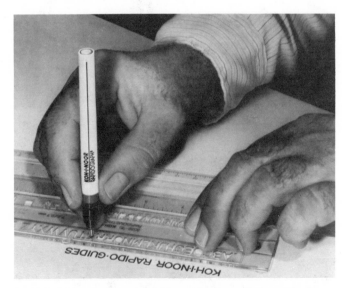

Fig. 10-24. Lettering should be added to orthographic drawings as neatly as possible. Templates are very helpful. (Koh-I-Noor Rapidograph, Inc.)

Fig. 10-25. Computer graphics programs contain commands for dimensioning drawings according to current professional standards. (Bausch & Lomb)

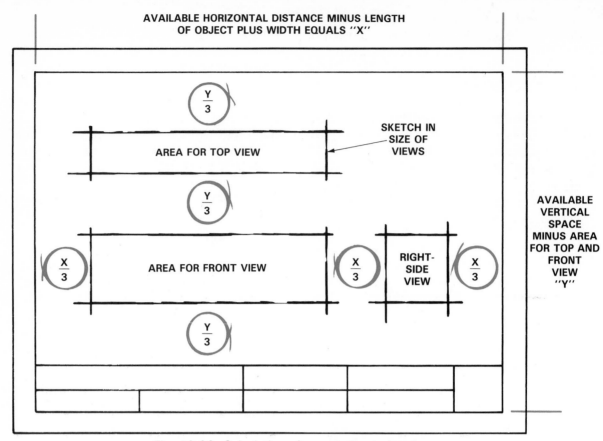

AVAILABLE HORIZONTAL DISTANCE MINUS LENGTH OF OBJECT PLUS WIDTH EQUALS "X"

$\frac{Y}{3}$

AREA FOR TOP VIEW

SKETCH IN SIZE OF VIEWS

$\frac{Y}{3}$

$\frac{X}{3}$ AREA FOR FRONT VIEW $\frac{X}{3}$ RIGHT-SIDE VIEW $\frac{X}{3}$

$\frac{Y}{3}$

AVAILABLE VERTICAL SPACE MINUS AREA FOR TOP AND FRONT VIEW "Y"

Fig. 10-26. Calculations for centering a drawing.

Fig. 10-27. Large sheets of drafting paper can hold many orthographic views on the same page.

smaller scale may be required to fit all views on the page. Once the desired scale is identified, the layout process can begin.

CENTERING DRAWINGS

Technical drawings look best when centered on the page. Remember, the center of an orthographic drawing is based on the dimensions of three separate views. Space between the views will further increase the total area. Therefore, the drafter must add many dimensions together to center the drawing. The method for calculating the center is given in Fig. 10-26.

If more than one set of drawings is to appear on the same sheet, layout is more difficult. This happens with many engineering and architectural drawings, Fig. 10-27. Generally, the largest (and most important) views are positioned first. Smaller, less important views are placed around the major details. Remember to maintain even spacing and borders for easier reading.

DEVELOPING ORTHOGRAPHIC VIEWS

After selecting the views to be drawn, layout of the drawing begins. Light guidelines are used to identify the location of each view, Fig. 10-28. On

a standard drafting board, use a T-square and triangles to develop all lines. It is best to start with the front view. The width and height lines developed in this view will help in starting other views. The general outlines must be constructed for each view. The location of holes, arcs, slopes, and other features should be noted. This may even help when planning where dimensions will be placed.

All drawings must be developed in perfect alignment, according to the rules of orthographic projection. That means the heights and widths are equal in each view. In addition, details must be drawn in respect to each other in adjoining views, Fig. 10-29. Hidden features are shown as dashed lines and are aligned as well.

Once enough guidelines have been sketched, darken the drawing. Drafting by hand means a drawing pencil or ink will be used, Fig. 10-30. If the drawing is done by computer, a printer or plotter will be required. Many printers are capable of

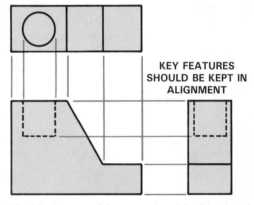

KEY FEATURES SHOULD BE KEPT IN ALIGNMENT

Fig. 10-29. Alignment of features is critical in developing orthographic projections.

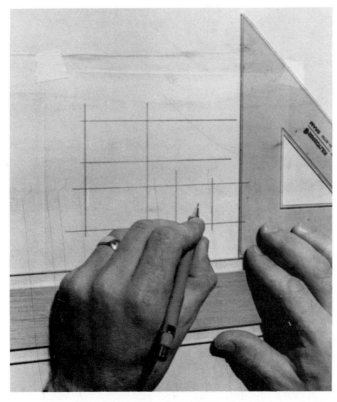

Fig. 10-28. Guidelines are useful in identifying the location of each orthographic view.

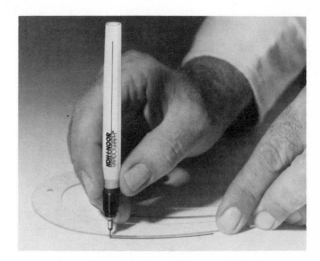

Fig. 10-30. Ink is used for many drawings because it reproduces (copies) much better than pencil. (Koh-I-Noor Rapidograph, Inc.)

Orthographic Drawing 99

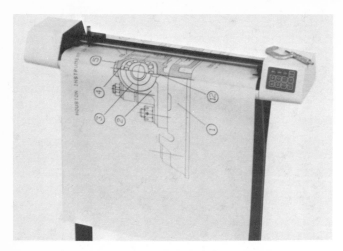

Fig. 10-31. Technical drawings are transferred from computers to paper with the aid of printers. (Houston Instrument)

making very finely detailed drawings, Fig. 10-31.

Lines should be drawn in order of their weight (boldness). Object lines are the darkest, boldest lines. Construct them first. Hidden and section lines are drawn slightly thinner. Lettering and dimension lines are usually done last.

Finally, the drawing should be cleaned up to improve its appearance. Guidelines should be erased carefully from around the drawing. Eraser shields (or simply a piece of paper) protect other lines and often prove useful. Do not disturb any lines immediately near a view or labeling. This may leave smudge marks on the drawing.

SUMMARY

Orthographic projection is called multiview drawing. The term orthographic refers to the method of projecting image lines from an object. A reference point is established at an infinite distance and perpendicular to an object. Image lines projected through a reference plane show how the technical view is developed.

Most orthographic drawings include three views. The front, top, and right-side views are most often used. Lines and details in each view are drawn in relationship to each other. All features are shown by either object (solid) or hidden (dashed) lines.

KEY WORDS

All the following words have been used in this chapter. Do you know their meanings?

Break lines, Convention, Cross-hatching, Dimension lines, Leaders, Multiview drawing, Orthographic projection, Point of reference, Reference plane, Symmetry, and Tolerance.

TEST YOUR KNOWLEDGE—Chapter 10

(Please do not write in this text. Place your answers on a separate sheet.)
1. ANSI and SAE are two examples of a _____.
2. _____ _____ is the process of projecting images from an object.
3. What three views are most often used in orthographic projection?
4. What lines are drawn the boldest in multiview drawings?
   a. Hidden lines.
   b. Centerlines.
   c. Dimension lines.
   d. Object lines.
   e. None of the above.
5. Cross-hatching is used to illustrate solid areas. True or False?
6. Name two orthographic conventions.
7. Which of the following is NOT a rule for using dimensions/notes on orthographic drawings?
   a. Place labeling on most descriptive features.
   b. Never label the exact size of a line, unless it is shown as full length.
   c. Never stagger dimensions.
   d. All of the above.
8. What technique is used to draw very long distances in a smaller space?

ACTIVITIES

1. Develop an orthographic drawing of this book. Supply front, top, and right-side views. If possible, develop this drawing on a computer. Use the same views.
2. Obtain a set of architectural plans for your school building. Make a list of all the details you recognize. Include those details you learned in earlier chapters. Explain what details are shown on each plan (floor plan, landscaping plan, etc.).
3. Develop an orthographic view of all parts of a ball point pen. Assemble these views onto a single sheet of drafting paper.
4. Do additional research on the process of orthographic drawing. Write a short report on what you learned. Include information on other uses for orthographic drawing (other than those discussed in the chapter).

11 | Pictorial Drawing

The information given in this chapter will enable you to:

○ *Understand drawing techniques for developing realistic pictorial drawings.*

○ *Describe the four forms of pictorial drawings used most often.*

○ *Identify oblique, isometric, technical, and perspective drawings.*

In Chapter 7, you learned that drawing is often divided into several classifications. Many drawings are developed using only two dimensions: height and width. Orthographic projections are examples of two-dimensional views. On the other hand, pictorial drawings are three-dimensional. These images have depth added to the views, Fig. 11-1. Texture and shading is sometimes added to make the drawing more realistic.

In most cases, pictorial drawings are designed to be as realistic as possible. This is especially true in the drawings produced by technical illustrators and architects. Designers in industry allow us to see products not yet made, Fig. 11-2. Their pictorial drawings often look like photographs. Architectural drafters provide a similar view of buildings in the design stage. Artistic renderings of new buildings permit others to see how the structure will appear when completed.

PICTORIAL DETAILING

When you look out a window, what do you see? Perhaps your attention focuses on items like trees and clouds. Maybe your attention is drawn toward cars, streets, and buildings. You might see colors and shapes. Sizes and distance will often become obvious to you. When you are developing pictorial

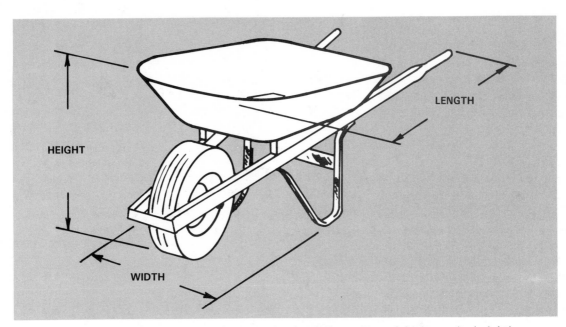

Fig. 11-1. Pictorial drawings illustrate the three dimensions of thickness (or height), width (or depth), and length.

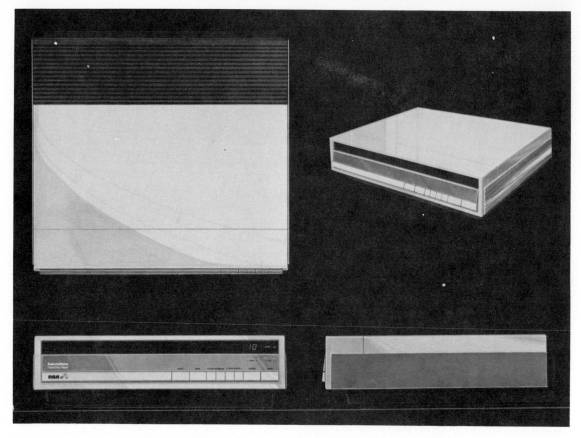

Fig. 11-2. A pictorial drawing of a video recorder. (RCA)

drawings, all of these details are important. These details make the drawing look realistic.

The person making a pictorial drawing must recognize many natural details. For example, size and detailing appear to decrease over distance. This is why train tracks seem to run together as they get farther away, Fig. 11-3. This is also the reason that trees, cars, and buildings appear smaller when they are viewed from a distance. When photographs are taken, these differences are simply recorded on the film. A drafter, however, must learn to draw these differences. There are several techniques used for reproducing these details. We will discuss four of these techniques.

FORESHORTENING

In nature, items appear smaller when observed from farther away. This is called FORESHORTENING, Fig. 11-4. It is the reason a mountain appears only inches high from a distance.

When drawing simple shapes, foreshortening is quite easy. Details developed towards the back of a view should be drawn slightly smaller, Fig. 11-5. Almost any item in the back of a picture should appear shorter by comparison to details in the front of a picture.

Fig. 11-3. To the eye, these railroad tracks seem to meet at the horizon. (Andy Johnston)

Fig. 11-4. Foreshortening can be seen when looking at a fence. The fence posts farthest from your eye appear shorter than boards closer to your viewing point.

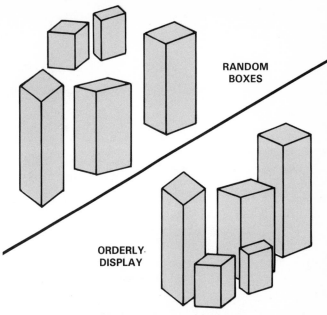

RANDOM BOXES

ORDERLY DISPLAY

Fig. 11-6. Overlapping creates organization of the objects in this view.

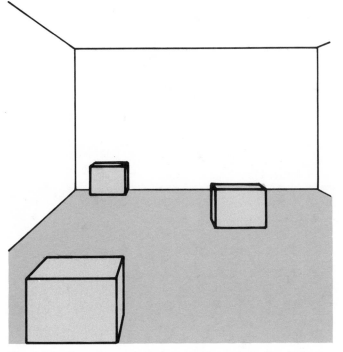

Fig. 11-5. Foreshortening sizes are based on the distance between the viewer and the object.

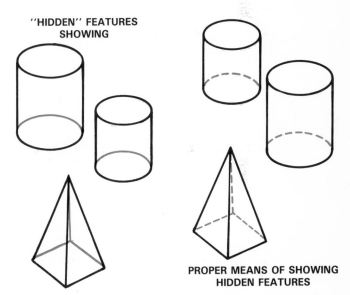

"HIDDEN" FEATURES SHOWING

PROPER MEANS OF SHOWING HIDDEN FEATURES

Fig. 11-7. When hidden lines are left in drawings, it is difficult to tell which items are located in front of others.

OVERLAPPING

The positioning of objects causes a condition we call OVERLAPPING. Items close to us are generally seen as whole, and close to actual size, Fig. 11-6. As objects are observed from greater distances, the viewer is at a disadvantage. Items may get in the way or "block" the view of other objects, Fig. 11-7.

This overlap is often ignored by drafters. They keep in lines that cannot be seen from their point of reference. Architects, however, often use overlapping to their advantage. Depth is easily shown by overlapping cars, streets, and trees in a drawing. Large hills in the background finish off a scene. Items that overlap help create a sense of distance or space.

HORIZONTAL VIEWING PLANE

We all have a point of reference (or viewing plane) at some level. Items positioned horizontally, in front of our eyes, appear on the same plane.

UPWARDS

STRAIGHT AHEAD

DOWNWARDS

Fig. 11-8. Several different viewing planes.

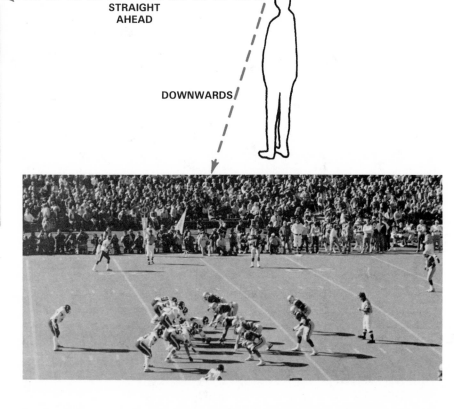

This plane is generally called EYE LEVEL. Objects above or below are seen in a different viewing plane, Fig. 11-8.

Most drawings developed by technical illustrators are done looking straight across at an object or scene. This means the image is projected in a horizontal viewing plane. Lines indicating height appear vertical (as always). Horizontal lines stretch from each side of the drawing and towards the rear of the scene.

Pictorial drawings can, however, be developed from a number of viewing planes. This is a major advantage over photographs. Large, permanent objects can be drawn as they would appear from "odd" angles. Perhaps you have heard of a "bird's

eye view?" This view is actually an overhead view. It is probably similar to what a flying bird might see from overhead, Fig. 11-9.

VANISHING POINTS

Humans can only see as far as the horizon, Fig. 11-10. In technical drawings, the horizon is actually represented by a set of points that form a line. Along this line are several spots called vanishing points. Most horizontal lines extend to one, or more, vanishing points.

Many pictorial drawings are started by first identifying several vanishing points, Fig. 11-11. Then lines of sight are developed. Simple shapes

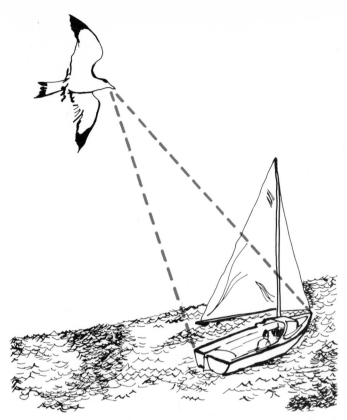

Fig. 11-9. Bird's eye views are developed from an overhead angle, looking down.

Fig. 11-10. The horizon is the farthest point visible to the human eye.

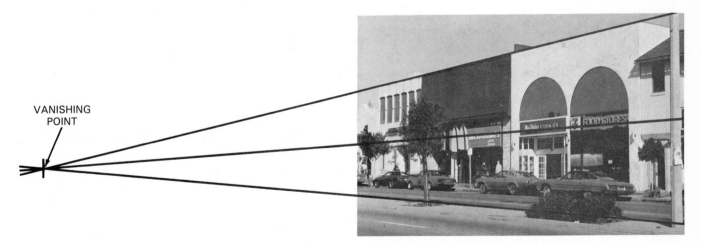

Fig. 11-11. Vanishing points are fairly easy to locate in photographs of any scene.

are sketched from these projection lines. Finally, object lines are produced with bold strokes. Shading and other details add to the drawing.

SHADING AND SHADOWING

All objects reflect light differently. Light is often projected onto a scene causing a variety of light and dark spots. We refer to this effect as SHADING AND SHADOWING, Fig. 11-12. The dark side of an object is said to be in the SHADE. On the other hand, the object itself may block rays of light. The darkened area caused by the blocked light is called a SHADOW.

There are several ways for shading a scene. One method starts by identifying the light source, Fig. 11-13. Light guidelines are then drawn to show how light strikes the object. Areas that receive little or no light are darkened. This allows drafters to identify surfaces requiring shading.

Pictorial Drawing 105

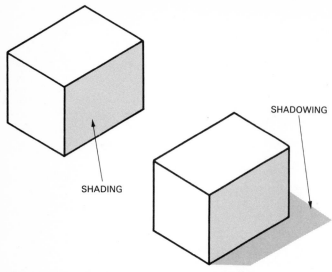

Fig. 11-12. Shading and shadowing added to a pictorial drawing.

SHADING

SHADOWING

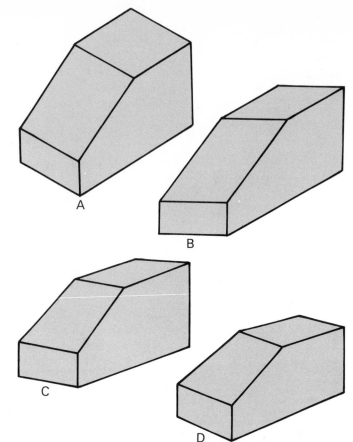

Fig. 11-14. Four types of pictorial views used most often in industry. A—Isometric drawing. B—Oblique drawing. C—Technical illustration. D—Perspective drawing.

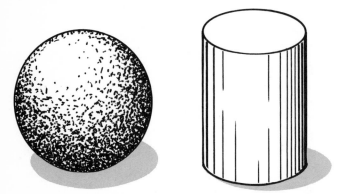

Fig. 11-13. A light source creates shadows and shaded areas. Most pictorial views have a light source coming from the upper left-hand corner of the drawing.

DEVELOPING PICTORIAL DRAWINGS

Industry uses four types of pictorial views for describing products or scenes, Fig. 11-14. Each of these types uses methods of projection to make the drawings. These techniques are fairly simple to use. Certain guidelines must be followed. These methods of projection are discussed in the following paragraphs.

ISOMETRIC AND OBLIQUE DRAWINGS

Two simplified methods of creating pictorial views are isometric projection and oblique projection. Oblique drawings are the easiest to produce, Fig. 11-15. However, they are not always accurate representations. Isometric drawings are more popular because the image is realistic. Actual sizes and shapes can be displayed easily, Fig. 11-16.

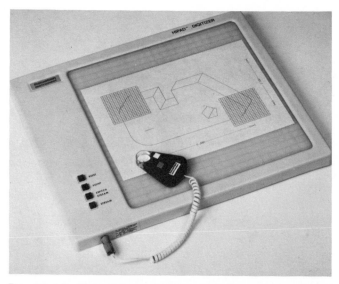

Fig. 11-15. This oblique drawing is being made with the help of a computer graphics pad. (Houston Instrument)

Oblique views contain a front view that is parallel to the picture plane. Depth is added by extending lines back at any angle. However, 30 degree angles are used most often.

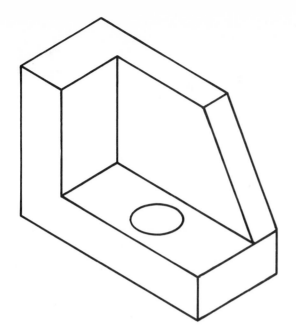

Fig. 11-16. Isometric drawings are clear and easy to understand.

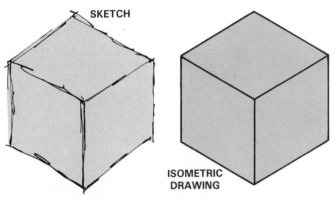

SKETCH

ISOMETRIC DRAWING

Fig. 11-17. With isometric drawings, lay out the outline of the drawing only, before starting on a final drawing. This will make centering the drawing on the sheet easier.

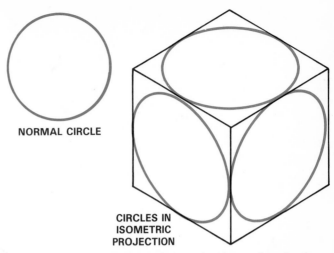

NORMAL CIRCLE

CIRCLES IN ISOMETRIC PROJECTION

Fig. 11-18. Circles in an isometric plane of projection.

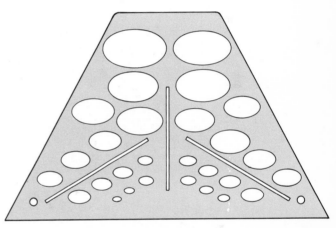

Fig. 11-19. An isometric template is used to draw ellipses for a final drawing.

In contrast, isometric views are more difficult to lay out. They are shown as they appear to the eye. Vertical edges are shown as vertical lines. Depth is shown with lines extending back to the left and right at 30 degree angles. Isometric views often become quite tall due to this method of layout. The view may need to be moved down the page in order for it to be centered. The total height cannot be larger than the available space on the sheet, Fig. 11-17.

Isometric circles are often confusing to draw. The best way to construct the rounded features is to "picture" which side of an object is involved, Fig. 11-18. An ELLIPSE will be created instead of a true circle. The diameter lines of the circle may be sketched to show the general shape. For the final drawing, an ISOMETRIC TEMPLATE should be used, Fig. 11-19. This permits the construction of clean, neat ellipses.

TECHNICAL ILLUSTRATIONS

Many types of pictorial drawings are used in modern industry. The most realistic are called TECHNICAL ILLUSTRATIONS, Fig. 11-20. These views often resemble photographs. Other types of technical illustrations are EXPLODED VIEWS and CUTAWAY VIEWS. Exploded views provide a way to illustrate the assembly of a product. Cutaway views are designed to show internal details the eye is not able to see.

Industrial drafters can describe many features of an object with a cutaway view. Cutaway views provide a "look" inside an object, Fig. 11-21. Section lines are often used to outline internal shapes. This shows the solid areas and the hollow areas.

Most cutaway views are made as an isometric

Fig. 11-20. This illustration of a satellite in orbit is a scene that would be difficult to photograph. (Ford Aerospace & Communications Corp.)

drawing. The entire object is lightly drawn to begin the view. The desired plane of reference is also shown on this view. Only items that would normally be seen are darkened. This creates the cutaway view, Fig. 11-22.

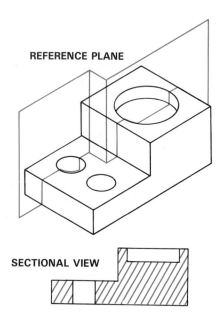

Fig. 11-22. A reference plane is used to identify the line where the object will be "cut." The exposed area that results from the cut is then sectioned.

AN OMNIDIRECTIONAL MICROPHONE

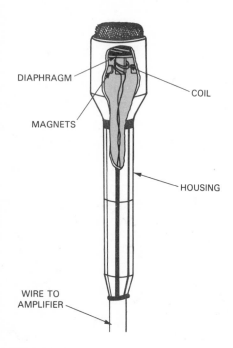

Fig. 11-21. A typical cutaway view.

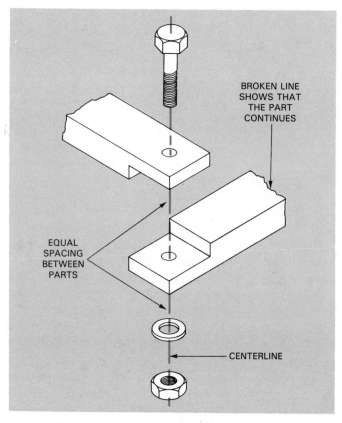

Fig. 11-23. Creating an exploded view.

Exploded views are constructed in much the same manner. A rough isometric sketch is prepared to align and space each object, Fig. 11-23. Spacing is sometimes difficult in exploded views. The total area required for the drawing is often very large. Individual parts are spread out around the page.

With exploded views, the centerlines must be drawn through each object to avoid confusion. These lines are also useful in showing how parts are to be positioned during assembly.

PERSPECTIVE DRAWINGS

Perspective drawings are the most accurate views for showing objects. This is due to the use of vanishing points and reference planes. Items are drawn as they appear to the eye. Lines extend to the horizon rather than at 30 degree angles. Refer to Fig. 11-24.

To understand perspective views, start by examining several photographs, Fig. 11-25. The image lines all seem to meet along several points on the horizon. These are the vanishing points.

To develop a perspective view, a horizon line is made at normal eye level, Fig. 11-26. Vanishing points are then placed on the horizon. All horizontal lines will run towards these points. Guidelines should be constructed from the front surface (of the object) to the vanishing points. Try to spread the points out for a realistic image, Fig. 11-27.

ISOMETRIC VIEWS OFTEN APPEAR AS "BLOCKS"

PERSPECTIVE VIEWS SHOW OBJECTS AS THEY APPEAR TO THE EYE

Fig. 11-24. Perspective drawings provide a more realistic view than isometric drawings.

VANISHING POINT

EYE LEVEL

Fig. 11-25. The image lines show where the vanishing points are in this photograph.

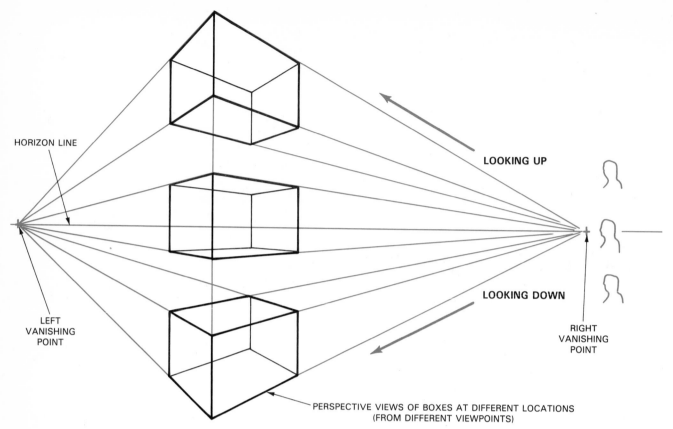

HORIZON LINE

LOOKING UP

LEFT VANISHING POINT

LOOKING DOWN

RIGHT VANISHING POINT

PERSPECTIVE VIEWS OF BOXES AT DIFFERENT LOCATIONS
(FROM DIFFERENT VIEWPOINTS)

Fig. 11-26. Developing a perspective view.

Fig. 11-27. Attention to detail is very important when making perspective views. Shapes and proportions must be realistic.

SUMMARY

Pictorial drawings are three-dimensional. They are more realistic than orthographic drawings. Oftentimes pictorial drawings look similar to photographs.

In order to get such a realistic drawing, the person developing a pictorial view must learn to use foreshortening and overlapping. The illustrator must also learn to recognize horizontal viewing planes, vanishing points, and shading and shadowing. All of these techniques help recreate the detailing of objects and scenes.

Pictorial drawings most often used in industry fall under four headings: isometric, oblique, technical, and perspective. Of these four areas, perspective is the most lifelike, since it is drawn as the item appears to the eye.

KEY WORDS

All the following words have been used in this chapter. Do you know their meanings?

Bird's eye view, Cutaway view, Ellipse, Exploded view, Eye level, Foreshortening, Horizontal viewing plane, Isometric template, Overlapping, Technical illustration, and Vanishing point.

TEST YOUR KNOWLEDGE—Chapter 11

(Please do not write in this text. Place your answers on a separate sheet.)

1. A skyscraper appears only a few inches tall when viewed from a great distance. This is an example of:
   a. Shortening.
   b. Overlapping.
   c. Foreshortening.
   d. None of the above.
2. Architects use _____ to show the depth of a view.
3. Explain the difference between eye level and a bird's eye view.
4. _____ _____ are a set of points along the horizon.
5. A shadow is:
   a. The dark side of an object.
   b. A darkened area caused by blocked light.
   c. A bright spot.
   d. None of the above.
6. Oblique drawings are more realistic than isometric drawings. True or False?
7. In perspective views items are drawn as they appear:
   a. From above.
   b. From below.
   c. To the eye.
   d. None of the above.

ACTIVITIES

1. Sketch a perspective view of the front of your classroom while seated at your desk. Then develop another view from the front of the classroom looking back at your desk.
2. Develop an exploded view of a large cabinet, with the top, sides, back, and doors removed.
3. Using a cereal box, develop isometric, oblique, and perspective drawings. Use a different view for each drawing.
4. Invite several architects to visit your class. Ask them to discuss the three types of pictorial drawings that were explained in the chapter. Ask them to include renderings of buildings they have designed that are done as pictorial drawings.

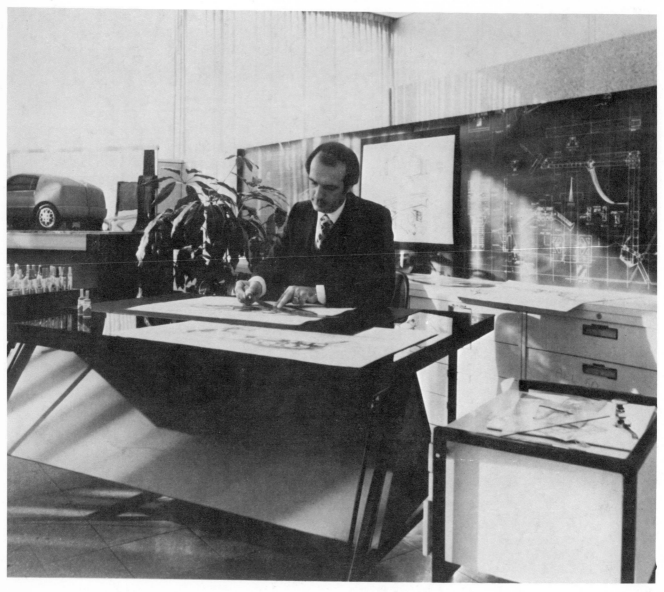

This drafter's skill allows him to communicate using drawings. (General Motors Corp.)

12 | Introduction To Graphic Communication

The information given in this chapter will enable you to:
- *Describe the development of graphic communication methods.*
- *Identify the major types of printing techniques used in graphic communication.*
- *Cite the uses of printing procedures in consumer and industrial products.*

Graphic communication is one of the industrial communication systems. Individuals and businesses both use this technology. In graphic communication all messages are transmitted (sent) using printed images. We see and use examples of printed materials daily. Items like books, magazines, and posters are produced by graphic means. So are beverage containers, brochures, and T-shirts, Fig. 12-1. Many forms of printing exist today. Communicating with graphic messages is an important practice in modern society. This chapter will introduce the various types and uses of graphic techniques.

HISTORY OF GRAPHIC COMMUNICATION

Printed images were first made in caves by our earliest ancestors. As time passed, people found new ways to communicate with graphics. We described several of these practices in Chapter 2 and Chapter 7. This chapter will focus on the reproduction (copying) of graphic images. Reproducing work includes everything from a single photograph to thousands of newspapers.

Early reproductions were done by hand. This obviously took a great deal of time. Just think how long it would take you to recopy this entire book. Handwriting is not an efficient method of reproduction. A faster method of copying materials was developed in China, then improved by the Europeans. An image (picture) was cut into a wooden block, Fig. 12-2. These blocks were used to print images on plaster and textiles.

The use of the printing press grew rapidly. Simple presses used in making wine and paper were

Fig. 12-1. Printing of graphic images is commonly used in packaging most products. Can you think of other items that use printing? (Graphic Arts Technical Foundation)

Fig. 12-2. Relief printing is done from a raised surface. Image areas (letters or symbols) are above the plate or printing block.

easily modified. An early printing press is shown in Fig. 12-3. Several improvements came from the introduction of the printing press. The greatest improvement was with the quality of impressions (transferred images). Locking the blocks in place prevented smudges and alignment problems.

Additional advances helped make printing more efficient. For example, it took hours to carve letters (in reverse) into printing blocks. With Johann Gutenberg's invention of movable type in 1450, text was easily assembled from racks of letters. This development led to an increase in the amount of printed material. Books became readily available for education and entertainment.

While movable type and the printing press vastly improved graphic communication, these methods remained quite slow. It would often take an entire day to set a page of type. Inventions like the typewriter and typecasting machine helped in

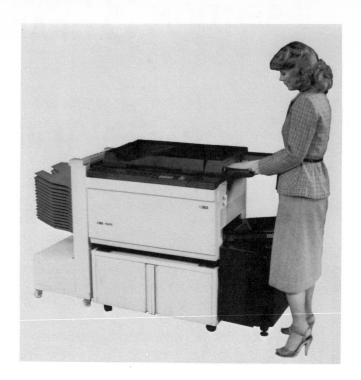

Fig. 12-4. Paper is usually manufactured in large rolls. Each roll may weigh over a ton. (Hammermill Paper Company)

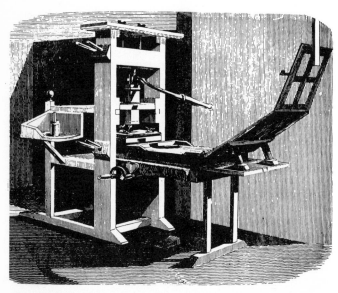

Fig. 12-3. Early printing presses were operated by hand. Paper was forced against inked blocks of type.

making printing more efficient. Steam-powered platen presses improved speed. Rotary presses were developed to print on both sides of paper at once. Rolled paper made the printing of newspaper much faster, Fig. 12-4. The paper is fed into these presses at high speeds.

At the same time, changes in other areas progressed. Photographic techniques developed steadily. New types of films improved the quality of photographs. Cameras with advanced features simplified the photography process. In addition, new types of printing evolved. A good example is LITHOGRAPHY or OFFSET PRINTING. This procedure permits the printing of images

from flat surfaces. SCREEN PROCESS PRINTING works by forcing ink through prepared screens. ELECTROSTATIC COPIERS use powdered inks for copying. Modern ink jet copiers show great promise, Fig. 12-5.

Fig. 12-5. Modern photocopy machines are capable of printing high quality images. This model creates graphic images by electrostatic means. (A.B. Dick Company)

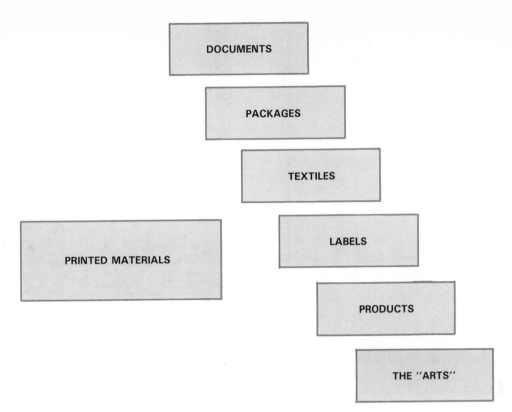

DOCUMENTS

PACKAGES

TEXTILES

LABELS

PRINTED MATERIALS

PRODUCTS

THE "ARTS"

12-6. The six groups of printed products.

GRAPHIC COMMUNICATION USES

Look around. Where are you reading this book? What forms of graphic communication do you see? The items you notice are produced by some printing technique. These techniques can be divided into categories. These categories include documents, packages, textiles, labels, products, and the arts. Each category is shown in Fig. 12-6.

This book is an example of a document. Printed documents usually require reading. They contain information for use at a later date. Newspapers, menus, maps, and brochures are documents. So are catalogs, greeting cards, and stationery. These materials contain printed images that add value and meaning.

Most packages use graphic communication. Designs and information are printed on cans, boxes, and plastic bags, Fig. 12-7. We read many packages for information and enjoyment. Have you ever read a cereal box at the breakfast table? Important details are given about the food inside.

Textiles are another major category of graphic communication. Many articles of clothing have printed messages on them. Messages on shirts show our favorite school, activity, or athletic team, Fig. 12-8. Designer labels make fashion items popular. Graphic messages on caps and uniforms display name-brands.

Labels are found everywhere in our society. People collect bumper stickers and decals. Address stickers are commonly used on envelopes. Many purchased products are marked by labels. These describe their contents, Fig. 12-9. Most labels are

Fig. 12-7. These aluminum foil packages are printed, folded, and sealed by machine. (Reynolds Metals Co.)

Fig. 12-8. T-shirts feature many types of messages school names, products, events, and clever sayings.

Printing practices are often used in the display of art. Painting is a form of artwork. To reproduce the work of artists, we copy the original paintings. Then posters and reprints of the drawings or paintings are produced. Photographs are also a form of artwork. Printing techniques are used to turn a negative into a usable photograph.

As you can see, printing techniques are used daily. Industry also relies on these techniques for routine operations. Activities such as recordkeeping and correspondence are examples. From com-

Fig. 12-9. These boxes of paper are ready for shipping. Notice the information on each label. (Hammermill Paper Company)

prepared by screen process printing.

Many consumer products include graphic images in their design. Board games and playing cards are good examples, Fig. 12-10. The printed designs create the required message or symbol. Other items are produced entirely by graphic means. For instance, plywood paneling is a printed product. Large presses are used to print a wood grain pattern on each panel. Circuit boards are produced in a similar way.

PLAYING CARDS
CIRCUIT BOARDS
WALLPAPER
MOVIE TICKETS
PLYWOOD
CALENDARS

Fig. 12-10. How many of these printed products do you recognize?

pany reports to employee paychecks, printed materials are vital in today's business world.

PRINTING TECHNIQUES

Today's technology has produced six major types of graphic processes. These include relief, screen process, continuous tone photography, intaglio, electrostatic, and lithography. They are shown in Fig. 12-11. We will examine these techniques, along with several specialty processes.

RELIEF PRINTING

Relief printing is the transfer of images from a raised surface. Typically, this surface is printed type (letters and numbers) contained on engraved blocks. The text can be hand or machine set. All relief work is prepared in reverse, Fig. 12-12. When impressed into paper, the image becomes readable.

Inks for relief printing are applied on the type with a roller. Then paper is brought in contact with the image area. A printing press is often used for this. Simple platen presses use the principles of relief printing, Fig. 12-13.

Relief printing is rarely used in today's industries. Relief work now mostly involves printing small orders. Wedding invitations are completed in this manner. Rubber stamps are often prepared using a special type of relief printing.

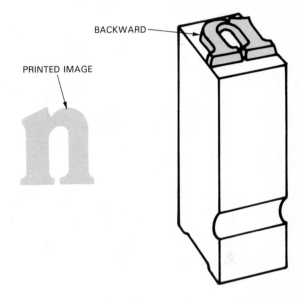

Fig. 12-12. Text for relief printing is created backwards or "wrong-reading."

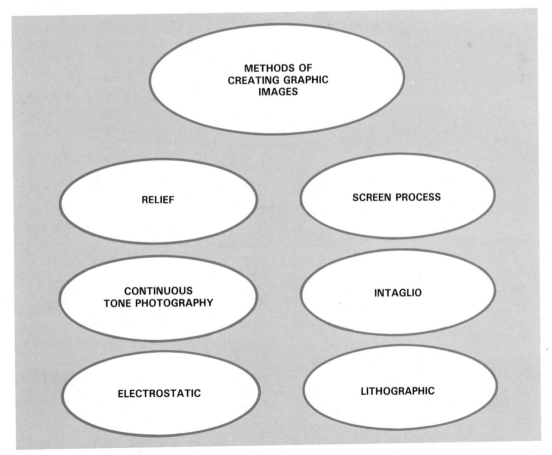

Fig. 12-11. The six methods of creating graphic images.

Fig. 12-13. The platen press is an early example of printing machinery. Have you seen a press like this before?

Fig. 12-14. Screen process printing is a common industrial practice. These items have been screen printed during production.

SCREEN PROCESS PRINTING

Screen printing starts with an original design. A stencil of the image is produced. Stencils are cut (in a film) by hand. (Industrial firms use photographic means to complete this task.) The stencil is then attached to a screen. Most screens are plastic or metal. At one time, all screens were silk. Consequently this procedure is often called SILK SCREENING. Today silk is seldom used because of its high cost.

Images are created by forcing ink through a screen with a rubber squeegee. The ink appears on the transfer medium. Separate screens are prepared for each color desired.

Forms of screen printing have been around since the middle ages. However, screen printing did not become widely used until recently. Modern industries rely heavily on screen process printing. Many common products are completed with screen images. Examples include T-shirts, signs, and glass containers, Fig. 12-14.

CONTINUOUS TONE PHOTOGRAPHY

Photography has two uses in graphic communication. Photographs may serve as a final product or they can help create pictures (or masters) to be used with other printing processes. Making masters requires additional steps, however. For example, stencils used in screen printing are often prepared by photographic means.

To produce continuous tone photographs, a camera and film are needed. Light entering the camera strikes the sensitive film. The film is later developed in a chemical bath. Resulting pictures may be black and white or in full color. Slides, prints, and movie film are produced this way. We will look more closely at this procedure in Chapter 16.

INTAGLIO

Intaglio (in-'tal-yō) printing transfers ink from an image engraved into a material to another surface. In industry this engraving is done on a metal plate or cylinder. Ink is applied to the engraved surface. A blade is used to wipe off excess ink from the surface of the plate. When paper is pressed onto the surface, ink lifts from the image area.

In industry this method is better known as GRAVURE PRINTING. This printing method is frequently used to publish newspaper supplements, and to make the designs on wooden wall paneling. The intaglio process is illustrated in Fig. 12-15.

ELECTROSTATIC COPYING

Much of modern office communication is reproduced using electrostatic copying. This is better known as PHOTOCOPYING. Xerox Corporation is a leading producer of electrostatic copying machines.

The electrostatic process relies on the principles of physics. From science, we know unlike charges (negative and positive) attract each other. In photocopying machines, a picture is taken on a metal plate. This picture is positive (+) and attracts a negative (−) toner. As paper passes the plate, toner is transferred to the paper. After heating, the powdered toner fuses to the paper. This leaves a permanent image on the page. Electrostatic copying is shown in Fig. 12-16.

Electrostatic copying is useful for other graphics work also. It is used to quickly enlarge or reduce artwork for design activities. The reproduction of personal documents (birth certificates, diplomas, etc.) is a large business also. These documents can be put in a personal file.

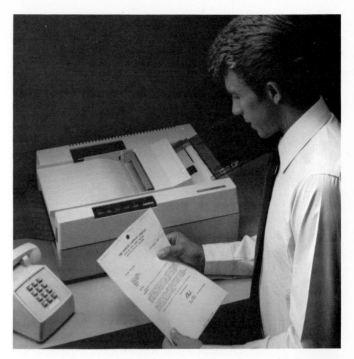

Fig. 12-16. An electrostatic printer creates graphic images almost by "magic." Electrical charges direct the flow of powdered or liquid inks.

THE GRAVURE PROCESS

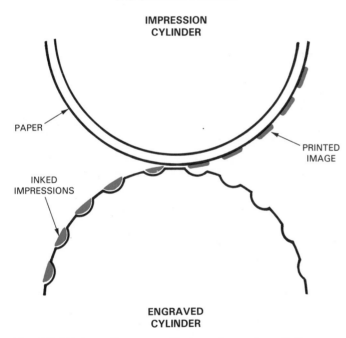

Fig. 12-15. Intaglio means "below the surface." Gravure (or intaglio) images are printed from inked areas below a smooth surface.

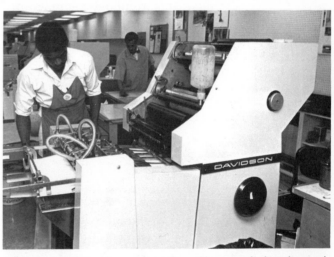

Fig. 12-17. Offset presses are found in many industries and schools. Flyers, pamphlets, and other handouts are often offset printed. (AFT—Davidson)

LITHOGRAPHY

Lithography, also known as OFFSET PRINTING, is printing from a flat surface. In offset work, an original is prepared on a metal, paper, or plastic plate. Normally, the plate is prepared chemically and mechanically to hold moisture. These plates are used in offset presses, Fig. 12-17.

The offset technique works on the principle that grease and water do not mix. The image to be printed is made with a greasy ink on the printing plate. The plate is coated with a film of moisture. The water sticks only to the clear areas (those without ink). Then ink is rolled over the plate. The ink sticks only to the image area. The inked image is transferred to a rubber blanket in reverse (offset). The blanket transfers the image to paper as the paper passes under the roller. The impression leaves a deposit of ink on the page.

Lithography is the most common industrial

printing process in use today. Quick printing establishments use this method. Larger firms also rely on offset procedures. Most books and magazines are printed by lithographic means. We will explore lithographic printing more in Chapter 15.

SPECIALTY PROCESSES

Recent technical growth has permitted new methods of printing. Some are used in schools and offices. Others are used in industry.

The DITTO PRINTING PROCESS works by typing or drawing onto a special master, Fig. 12-18. This leaves a carbon deposit on the carrier sheet. A spirit fluid softens the carbon during the printing process. This allows the images to be deposited on paper. The color of the print is determined by the master. A purple image is most common. Green, red, black, and blue are also available.

The MIMEOGRAPH PRINTING PROCESS machine is another specialty process. Its masters can also be prepared by typing or drawing. This procedure is similar to screen process printing. The master acts like a stencil. This stencil is attached to a cylinder on a mimeograph machine. Ink is forced through the openings of the master, Fig. 12-19. As paper passes the master, a deposit of ink is transferred. Most work is done with black ink.

Fig. 12-19. The mimeograph process resembles screen process printing. Can you describe how the two are alike?

DIAZO is another specialty printing process. It has many uses in business and industry. The most common is to reproduce engineering drawings. The resulting copies are called WHITE-PRINTS. Unlike blueprints, the background is white and all lines are blue. Making whiteprints is relatively easy. The original is passed through a diazo machine along with a copy paper. The image is copied by exposing the paper to light and ammonia gases.

Heat is also used to reproduce specialty products. It is especially popular in T-shirt printing. Designs are created using offset or screen process printing. The designs are printed on thin tissue paper. The images are lifted from the carrier sheet to the textile surface. We usually call this process HEAT TRANSFER PRINTING.

Another thermal technique is used in small electronic calculators. The paper on these instruments is heat sensitive. Wires (or a stylus) contact the rolled paper. The dye in the sheets creates the imagery.

A final specialty process is INK JET PRINTING. In this method, tiny drops of ink are shot onto paper. A rectangular pattern of dots form characters. This procedure is usually controlled by computers. In fact, the most common use of this practice is computer printouts. Mailing labels and name tags are often produced by ink jet.

Fig. 12-18. The ditto machine is among the most commonly used printing devices today. Many school materials are completed using ditto equipment.

SUMMARY

Graphic communication is one of the industrial communication systems. All messages transmitted are printed images. Documents, packages, textiles, labels, products, and artistic work are all types of graphic communication. They are all produced using some type of printing technique.

Graphic communication has grown and improved steadily throughout history. Inventions like the printing press and movable type helped this growth. Today there are six major types of graphic processes. These processes have improved the quality of printed images.

KEY WORDS

All the following words have been used in this chapter. Do you know their meanings?

Continuous tone photography, Diazo, Ditto, Electrostatic copying, Gravure printing, Heat transfer, Ink jet printing, Intaglio, Lithography, Mimeograph, Offset printing, Photocopying, Relief printing, Screen process printing, and Whiteprints.

TEST YOUR KNOWLEDGE — Chapter 12

(Please do not write in this text. Place your answers on a separate sheet.)

1. Graphic communication sends all its messages using _____ _____ .
2. Give five examples in which industry uses graphic communication to carry out daily operations.

MATCHING QUESTIONS: Match the definition in the left-hand column with the correct term in the right-hand column.

3. __ Projecting images through film onto light sensitive paper.
4. __ A popular office reproduction method.
5. __ Based on the principle that ink and water do not mix.
6. __ Thermal method used to print images on textiles.
7. __ Printing from a raised surface.
8. __ Printing from an engraved surface.

a. Relief printing.
b. Screen process printing.
c. Continuous tone photography.
d. Intaglio printing.
e. Electrostatic printing.
f. Lithographic printing.
g. Ink jet printing.

9. __ Computer directs drops of ink into medium.
10. __ Forcing ink through a stencil.

h. Heat transfer printing.

11. Which printing method would be used to reproduce a minimum number of office copies?
   a. Relief.
   b. Screen process.
   c. Offset.
   d. Electrostatic.
12. Which method of printing would be used to print wall paneling?
   a. Relief.
   b. Screen process.
   c. Intaglio.
   d. Offset.
13. Photocopying newspapers and magazines for future use is an example of what part of the communication process?
   a. Designing.
   b. Coding.
   c. Transmitting.
   d. Storing.

ACTIVITIES

1. Collect samples of products that use graphic techniques in their manufacture. Identify the type of printing practice used in producing these items. Which items are your favorites? Why?
2. Research the development of graphic communication. Construct a time line to illustrate this development.
3. View movies on various methods of graphic reproduction.
4. Have a professional photographer visit your class to explain the techniques used to develop and print photographic film.
5. Use an electrostatic copier to create a class newsletter. Include short articles, photographs, and artwork done by you and your classmates.
6. Research the career of Johann Gutenberg. Present the information to your class as an oral report.
7. Arrange a visit to a quick print shop. Observe the printing methods used.
8. Create a display board featuring the six major types of graphic techniques.

13 | Graphic Design and Production

The information given in this chapter will enable you to:
○ *Describe the stages used to create and transmit graphic messages.*
○ *Explain common production techniques in developing a graphic message.*
○ *Compose a graphic message using the elements and principles of design, and graphic production techniques.*

The purpose of a graphic design is to transmit a message. The message may instruct, persuade, inform, or entertain the audience. It should look pleasing and be easy to read. If the design is not good, the message will not be understood.

In order to design effectively, certain design elements and principles must be understood. These are then used in the production of the graphic design.

DESIGNING

Designing a graphic message is the first step toward a finished product. A design is decided upon before photographs are picked or text is typeset.

Every graphic message has a purpose and an audience. These factors are combined with creative ideas to create the message.

DESIGN ELEMENTS

To create a graphic message requires following the ELEMENTS OF DESIGN. Shape, mass, texture, lines, and color are design elements. These elements are also considered when constructing artwork and copy. The design phase begins with development of basic lines, Fig. 13-1.

LINES are strokes made with pens, pencils, or tape. Lines vary in width and length, Fig. 13-2.

Fig. 13-1. Lines define space and give shape to objects. Simple sketches like these are called "doodles." (RCA)

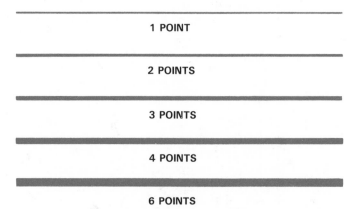

1 POINT

2 POINTS

3 POINTS

4 POINTS

6 POINTS

Fig. 13-2. The width (thickness) of a line is indicated by "points." Point values are shown here.

Lines can be straight, wavy, or curved to help create the desired graphic effect. Notice, too, that words and handwriting are made of lines.

MASS refers to the amount of space taken up. Dark (or bold) objects appear larger. That is also why bold print is used in many books. The bold type appears larger and more important than normal type. This is a key design element.

TEXTURE usually describes the amount of roughness or smoothness on an object. The quality of surfaces affects what we see or feel. Shading on drawings creates a feeling of texture. It provides realism, Fig. 13-3. Graphic designers realize this and use this element.

SHAPE is a combination of lines and mass. Examples include rectangles, circles, and other geometric designs. You may recognize the familiar shapes of the signs in Fig. 13-4. Many shapes add form or structure to a message. The shape of lettering also creates different impressions.

The final element of design is color. COLOR adds emphasis to graphic work. Red and yellow attract attention. Blue and green are calming (or mild) colors. Changing the colors of wording draws attention to the printed material.

DESIGN PRINCIPLES

Design principles describe the nature of the layout. Balance, contrast, rhythm, proportion, and unity are design principles. Three examples are shown in Fig. 13-5.

Fig. 13-3. Sketches can be shaded to closely resemble actual scenes.

Fig. 13-4. Signs and symbols are common graphic designs. A red sign or a yellow rectangle always transmit a particular message.

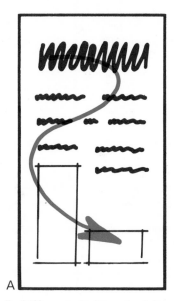
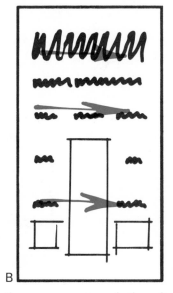
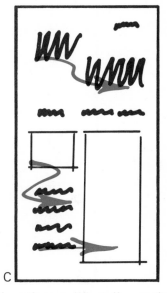

A

B

C

Fig. 13-5. Different design principles affect the way we see a design. A—Rhythm: the eye ''flows'' over the design. B—Formal balance: the eye follows the design from left to right. C—Contrast: the eye wanders across the design.

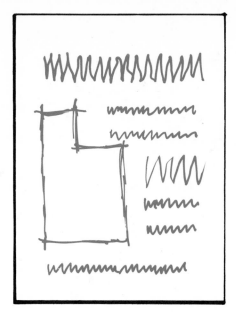 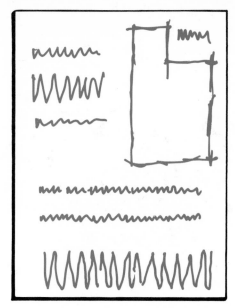

Fig. 13-6. Notice how balance affects the design of these simple layouts.

BALANCE deals with the location of parts within a layout. If the parts are centered, it is referred to as FORMAL BALANCE. Each item is orderly and evenly weighted. If the arrangement of parts is random, it is called INFORMAL BALANCE. Fig. 13-6 shows two examples of informal balance.

CONTRAST is important in providing a point of emphasis in a layout. It can be achieved with colors, text, or lines. Bold styles of lettering often provide contrast. Color or shading of artwork pro-vides contrast also. Any attempt to "catch" your eye is usually an example of this concept.

RHYTHM deals with the way a message is con-structed. Certain designs seem to guide your eyes along the message. The peaks in Fig. 13-7 achieve rhythm. They suggest a natural rhythm. The eye can easily read the text by following the artwork.

PROPORTION is the relationship of sizes in the design. Sizes of objects must appear uniform in the whole design. Titles must not be too much larger or smaller than the rest of the text. Large

Fig. 13-7. The design of this poster is improved with the addition of lines that create rhythm. Your eye tends to focus on the artwork and titles.

pictures often detract from the design. It is sometimes difficult to complete a pleasing design. Therefore, mathematical calculations are used to establish creative designs, Fig. 13-8.

UNITY is the final design principle. Its function is to pull the total design together. Designs that lack unity rarely communicate a message well.

Fig. 13-9. Computers are examples of modern technology. The term, "computer" looks odd in certain old-fashioned typefaces. Do you agree?

The exchange of ideas or feelings becomes confusing. This is evident in Fig. 13-9.

CODING

In graphic communication, after the message has been designed it is coded. In the case of a graphic message, coding is also known as layout.

LAYOUT is the assembly of copy (text) and artwork (illustrations), Fig. 13-10. The pages of this book are examples of layout. After the text was written and the illustrations selected, a formal layout was designed.

Fig. 13-8. Many illustrations contain titles or artwork that appear "out of place." What is wrong with this design?

Fig. 13-10. Graphic designers often create a rough layout of a job. They are then able to compare designs, and select the best format.

The copy includes the words, sentences, and paragraphs of the book. It also includes the captions that go along with the illustrations. The illustrations include drawings and photographs. They add meaning to the copy.

Layout starts with THUMBNAIL SKETCHES. These are small, crude drawings, Fig. 13-11. Thumbnail sketches (often called "doodles") are initial ideas for layout of the message. Thumbnail sketches are used as reference when discussing and developing a final design. When one idea is accepted, layout moves to the next stage.

In a rough layout, the idea is developed further. A ROUGH LAYOUT is more accurate and detailed than a thumbnail sketch. And it is produced at full size. This means text and artwork will be shown in their proper proportions, Fig. 13-12.

The rough layout provides a guide for a COMPREHENSIVE LAYOUT. It is used by the layout person as a guide during the reproduction work. Actual type and illustrations are still not used. Therefore, final corrections of the layout can still be made. An example of a comprehensive layout is shown in Fig. 13-13.

Next, the designer must create an assembly of the complete message, called a PASTEUP, Fig.

13-14. In this process, the copy is set in type, and the necessary artwork is located.

Copy can be set in type in several ways. Hand lettering and stencils are two procedures. If a more precise look is needed, a typewriter can be used to create the copy. Transfer lettering is also used in creating copy. Leroy and Kroy machines are also commonly used, Fig. 13-15. Machines, however, prepare most type set by commercial firms. Type is made photographically on light-sensitive film. This is called PHOTOTYPE-SETTING, Fig. 13-16.

The artwork (photographs, illustrations) usually comes from one of four major sources. They can be hand drawings, clip art, photographs, or mechanical drawings.

CLIP ART is drawn by professional artists, Fig. 13-17. It usually is sold in book form with several drawings on each page. The drawings are arranged and sized in several ways. This makes it possible to use them in different displays.

The pasteup is then placed on a clean, white sheet of paper or cardstock. Tracing paper or special layout sheets can also be used. Artwork and type must be located and positioned. Blue pencils are used in marking the pasteup sheets. These lines

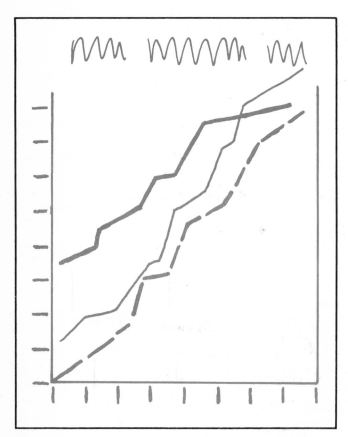

Fig. 13-11. Thumbnail sketches give the basic idea of a message. We create these simple drawings to graphically express ideas.

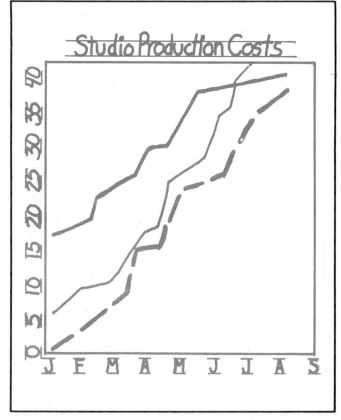

Fig. 13-12. Rough layouts are prepared to check the appearance of the final design. The next illustration shows how this graph is completed.

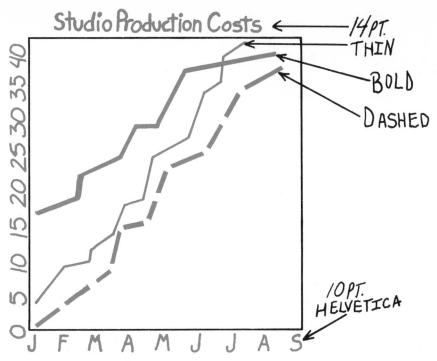

Fig. 13-13. A comprehensive layout is designed as a guide for the layout artist. Contrasting colors (black and white) permit copying with great ease and clarity.

Fig. 13-14. Each page of a newspaper is carefully pasted up by layout artists. The editor (standing) is checking for possible layout problems. (San Angelo Standard Times)

will not reproduce during processing. Artwork is secured with rubber cement, glue sticks, tape, or wax. The result is a neatly prepared MECHANICAL LAYOUT, Fig. 13-18. This layout is CAMERA READY and will serve as the master

Fig. 13-15. The Kroy and Leroy lettering machines are useful in making clear copy.

Fig. 13-16. Most phototypesetting is done on a computer. This editor types a story on the computer terminal, and a machine sets the type for the article. (San Angelo Standard Times)

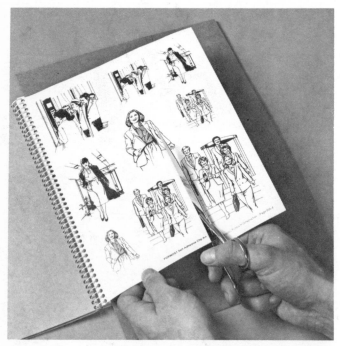

Fig. 13-17. Commercially prepared clip art is useful for design work. This artwork saves time in completing graphic layouts. (Graphic Products Corp.)

for future production work. It is an exact copy of the finished product.

TRANSMITTING

While designing and coding the message it is important to think about ways to reproduce (transmit) the message. We discussed the six methods of transmitting printed images in Chapter 12. Selection of the best method is often determined by purpose, time, and cost.

The purpose of the message sometimes determines the printing process used. If the purpose of the message is to create a strong, lasting impression, a visual display may be used. An offset process would reproduce this message. If the message is basically text, electrostatic copying or offset lithography are two options. Can you identify other ways an item might be printed?

The transmitting medium often determines the printing procedure. We know that T-shirts are best printed by screen process methods. Transfer images (applied with heat) might also be selected. However, the methods are somewhat limited to these options.

In industry, other restrictions limit the options. When printing on metal, offset or screen process methods are used. If a flat surface is available, offset printing is used. Round objects (cans and glasses) require screen process printing. In addition, different inks are needed for these special jobs.

Other factors determine why one process is better than another. Time and cost are critical. Equipment is also a key element in most printing firms. Experience of the printer may also be an important factor.

RECEIVING AND STORING

By selecting the transmitting method, we also identify the receiving medium. Magazines, clothing, and beverage cans are all printed by various

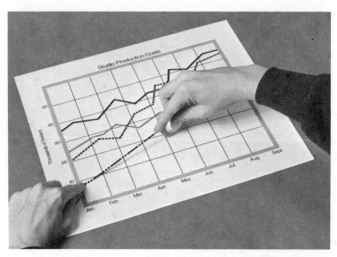

Fig. 13-18. The layout of graphic designs is simplified with some basic equipment and supplies. Does this artwork look neat and attractive? (Graphic Products Corp.)

128

methods. But they are all examples of receiving systems. The visual message remains on the product. In the case of computer generated charts, the receiving medium is paper, Fig. 13-19. Other receiving mediums include wood, textile, plastic, and metal.

Many types of storage mediums are available. Each has a different use. Photographic paper (from developed film) is a storage medium. The information used to make the chart in Fig. 13-19 is probably stored on a disk in a computer file. Many libraries store reference materials (newspapers and magazines) on microfilm.

The receiving and storing of graphic messages is important in communication systems. Photography is a common example. Many new methods of printing will be refined in the years ahead. This will add new work challenges for graphic designers.

SUMMARY

There are five stages a graphic message goes through in order to become a finished product. During the design stage, design elements and design principles are used. These help shape ideas for transmitting the message. This leads to the coding stage. It starts with thumbnail sketches of the ideas from the design stage. The thumbnail sketches are then developed into a rough layout, then a comprehensive layout, and finally a mechanical layout. The third stage is transmitting. At this time the message is actually reproduced. Receiving and storing are the two final stages.

KEY WORDS

All the following words have been used in this chapter. Do you know their meanings?

Balance, Camera ready, Clip art, Color, Comprehensive layout, Contrast, Design elements, Design principles, Formal layout, Informal layout, Layout, Lines, Mass, Mechanical layout, Pasteup, Proportion, Rhythm, Rough layout, Shape, Texture, Thumbnail sketch, and Unity.

TEST YOUR KNOWLEDGE — Chapter 13

(Please do not write in this text. Place your answers on a separate sheet.)
1. List the five stages in developing a graphic message.
2. Identify each of the following as either a design element or a design principle.
   a. Contrast.
   b. Shape.
   c. Mass.
   d. Lines.
   e. Unity.
   f. Balance.
3. _____ _____ are used to create a graphic message. _____ _____ describe the nature of the layout.
4. Explain the difference between formal balance

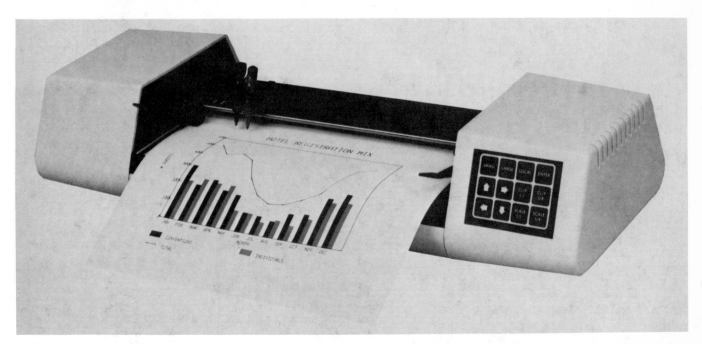

Fig. 13-19. This graph was drawn on standard bond paper. A plotter (computerized printer) quickly completed the drawing. (Houston Instruments)

and informal balance.

5. Layout is the assembly of artwork. True or False?

6. Which is (are) not a step(s) in the layout process?
    a. Mechanical layout.
    b. Computer layout.
    c. Thumbnail sketch.
    d. Rough layout.
    e. None of the above.

7. Define pasteup.

8. When selecting a method for transmitting a graphic message, _____ , _____ , and _____ often determine the best method.

9. Give five examples of storage for graphic messages.

ACTIVITIES

1. Collect several posters advertising different events. Identify the use of design elements and principles.

2. Create designs for school stationery. Develop the idea from thumbnail sketches through to a mechanical layout.

3. Use clip art and transfer lettering to design a schedule of school activities.

4. Use sketching techniques to develop a class logo or symbol.

5. Collect a variety of printed materials. Discuss the style of lettering used for titles, captions, and text.

6. Examine different types of paper and card-stock used in graphic communication.

This designer uses clip art to complete her layout. (Graphic Products Corp.)

14 Screen Process Printing

The information given in this chapter will enable you to:
○ *Describe the purposes and uses of screen process printing.*
○ *Prepare screen process stencils using a variety of methods.*
○ *Demonstrate skill in using screen process printing techniques by producing several products.*

Screen process printing is a popular method of graphic communication. It involves forcing ink or paint through a stencil. Typically, the stencil is attached to a framed screen. You will remember that early screens were made from silk. That is why the process is often called silk screening.

Screen process printing is a versatile graphic communication process. It is useful for many industrial and classroom purposes. Images can be reproduced on a variety of materials. Even very detailed prints are possible with stencils prepared using photographic film.

PURPOSES AND USES

Screen process printing is used when a heavy deposit of ink is required. The ink is usually applied to flat surfaces. It feels raised above the surface. Special machines permit printing of rounded messages on jars and cans.

Screen process printing is a specialty process. Very few products use paper as a transfer medium. Several exceptions include decals, stickers, and heat transfers. These are produced on thin sheets of tissue paper.

Most screen process work is done, however, on materials other than paper. Street signs and posters are examples, Fig. 14-1. Dials and labels on machines are also screened. These all have been printed using special machines. Other examples include electrical circuit boards and many types of clothing, Fig. 14-2.

There are several disadvantages with screen process printing. This method is slow as a reproduction process. One reason for this is the long

Fig. 14-1. Many visual displays are completed by screen process printing. Can you think of other examples besides those shown here?

Fig. 14-2. How many printed shirts, hats, and jackets do you own?

drying time required. In addition, it is difficult to transfer images having fine detail. Therefore, bold text would reproduce well, but photographs would not.

STENCIL PREPARATION

Many methods for preparing stencils have been developed. These include hand-cut paper and film stencils, photo direct and indirect stencils, and tusche stencils, Fig. 14-3. Only the hand-cut and photo indirect methods will be explained in this text. These are the primary methods used in schools.

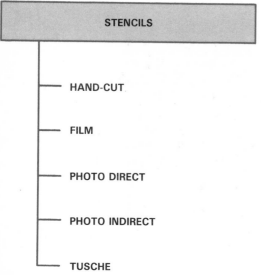

STENCILS

- HAND-CUT
- FILM
- PHOTO DIRECT
- PHOTO INDIRECT
- TUSCHE

Fig. 14-3. Methods of preparing stencils.

HAND-CUT PAPER STENCILS

Paper stencils for screen process printing are easy to prepare. This method is simple and inexpensive. However, three major drawbacks limit its use. First, objects with centers may not be printed, Fig. 14-4. The outside lines of the image area will print clearly. But the center part would fall out. The use of lettering is thus limited. All parts of the image must remain attached.

The second disadvantage is that the design must remain simple. Fine detailing is not possible with paper stencils. Images with sharp lines and curves are best. Various geometric shapes also work well.

The final drawback is the limited number of copies that may be reproduced. Paper stencils will not last very long. Printing inks and pressures break these stencils down quickly.

Preparation of paper stencils starts with an original design. The design is drawn on paper, Fig. 14-5. Kraft (brown wrapping) paper or grocery bags are good stencil papers. Waxed paper is also good for use with oil-based ink. The paper should be about the size of the screen frame. It should not be creased or torn.

The design is traced or drawn in the middle of the stencil paper. Tracing may be done on a light table. Lines should only be dark enough to be seen (not bold or heavy). Pencils are preferred for tracing.

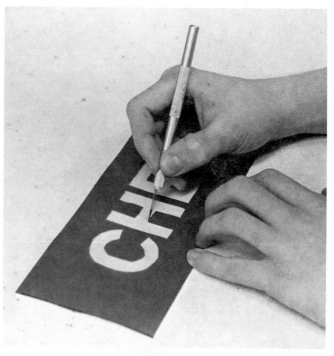
Fig. 14-4. Hand-cut stencil designs must be simple. Centers of complex figures are difficult to prepare. Avoid designs with internal features.

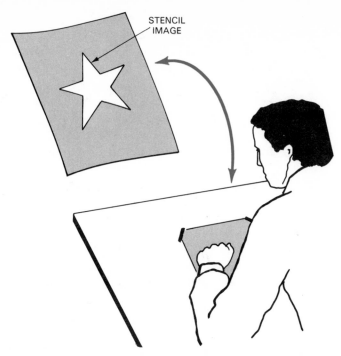

STENCIL IMAGE

Fig. 14-5. Simple designs are easily drawn directly on a hand-cut stencil.

When the design is transferred onto the paper, the stencil is then cut. A sharp stencil knife is best for cutting. Cut the stencil out on a hard surface. Pieces of press board or glass are excellent surfaces on which to cut. Do not use tabletops! Carefully cut out the stencil with smooth strokes, Fig. 14-6. Edges should not be jagged or torn. This can happen when using a dull knife.

Location of the stencil is very important. Each print made should be in the same place as the print made before it. In order to consistently position the transfer medium with the stencil, REGISTRATION MARKS are used.

Tape the original design on the printing base. Make certain it is in the correct position. It should look exactly like the completed product. Position it in relation to the frame. Next, set the guides.

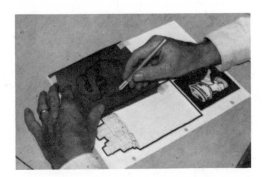

Fig. 14-6. A stencil knife is used to cut paper stencil. Follow the pattern carefully with the blade.

GUIDES mark the base in order to position each sheet in the same place. Masking tape is often used as a guide. The transfer medium is placed in position against the guides, Fig. 14-7.

The stencil is now ready for adhering (sticking) to the screen. Use several drops of white glue to secure the stencil. Spread the drops around the design area. Be sure no glue gets on the opened stencil section. This would keep ink from passing through the image area during printing. Finally, allow several minutes for the glue to dry.

HAND-CUT FILM STENCILS

Two types of hand-cut films are often used in screening work. They are aqua (water) and lacquer films, Fig. 14-8. Their preparation is similar to that used for paper stencils. The major difference is in adhering the film to the screen. Paper stencils are positioned and glued. Films are attached with water or special fluids. Aqua film is adhered to the screen with water. Lacquer or oil-based inks are used during printing. Lacquer-based film is attached with an adhering fluid. This special fluid does not dissolve the stencil as quickly as some other thinners. Water or oil-based inks are used to print with these stencils.

To prepare an aqua or lacquer film stencil, a design is needed. One master is used for each color to be printed. For example, to print a bull's-eye, three stencils are needed, Fig. 14-9. The inside circles (red and white) each require a screen. The thin, black lines (the outside of the circles and the bull's-eye) also need a screen.

Next, obtain the film to be used. Most films come in large rolls or sheets. However, a small

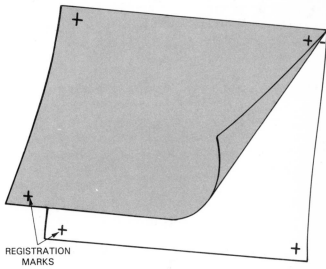

REGISTRATION MARKS

Fig. 14-7. Registration marks help in positioning the stencil for proper alignment.

Screen Process Printing   133

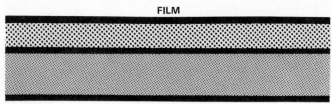

FILM

BACKING SHEET

Fig. 14-8. Hand-cut film has two layers. The film layer is cut when making the design. The backing sheet remains in place until the stencil is adhered to the screen.

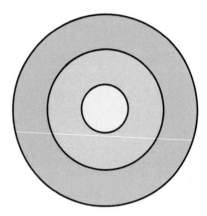

Fig. 14-9. When multiple colors are to be printed, separate screens are required. Do you know why?

piece of material is all that is needed for the stencil. Remember, each color requires a stencil. Generally, the film should be cut about two inches larger than the designed image. This allows extra working space during preparation and the printing process. Tape the design on a hard surface. Again, a press board or a light table are best. Lay the stencil film on top of the design. The shiny (or gelatin) side MUST BE FACING UP. The design should be in the middle of the film. Use tape to secure the film in its proper place, Fig. 14-10.

With a stencil knife, carefully cut the film. Trace the lines of the image area. Light, even strokes with a sharp blade work best. Do not press too hard. Only the gelatin coating is to be cut. The clear backing sheet of the film should remain intact. Cutting entirely through the backing will cause problems later. Lines forming corners may be overcut. These sharp cuts will fuse together when adhering to the film.

Once the lines are cut, slowly peel back the gelatin coating in the design area. The sharp tip of the knife is helpful in lifting the edges. Slowly remove the FILM ONLY inside the scribed lines. The area to be printed will become noticeable.

You are now ready to adhere the stencil to a screen. Use a screen that is clean and free of holes or ink. Position the screen over the stencil. Again,

registration marks may be needed to find the proper position for the stencil. Aqua film requires tap water to dissolve the film. Adhering fluid (or lacquer thinner) is used with lacquer film, Fig. 14-11. Moisten a rag with your adhering fluid. A second dry rag should be handy for cleanup purposes.

The proper adhering fluid will soften the gelatin layer on the film. It will actually dissolve the film into the screen. With the dampened rag, press the screen into the film. Then dab up the moisture with the dry rag. Repeat this process over the entire stencil, Fig. 14-12. Always rub lightly. Too much pressure will dissolve (or burn out) areas of the stencil. Take your time for this step. Rushing the procedure will probably ruin your stencil.

As the film dries, it will darken. If there are light spots on the film, an incorrect amount of fluid was used. These areas may not print properly. A new screened master may be needed. After the stencil is adhered, let the screen dry totally. Propping it in front of a fan will speed the process. Your hand-cut film stencil is now ready for printing.

PHOTO INDIRECT STENCILS

Paper and film stencils are prepared by hand cutting. Photo indirect stencils use light and chemicals to produce the stencil. As with hand-cut film stencils, separate originals are prepared for each color to be printed. The original design must be created in a solid, opaque color. This is usually black. Opaque films (used in industry) also come in red and orange.

After the original is created, a positive must be made. There are three ways to accomplish this, Fig. 14-13. A good, sharp photocopy is best for simple designs. A Thermo-fax transparency is made from the photocopy. The positive appears as a black image on clear film. A second way to

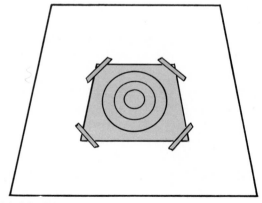

Fig. 14-10. Film is taped in place before printing. If the stencil should move, you may have to start the job over!

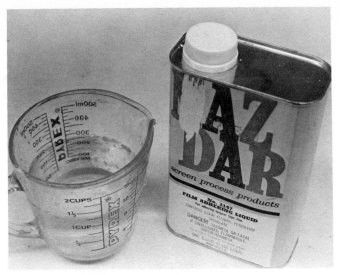
Fig. 14-11. Adhering fluids.

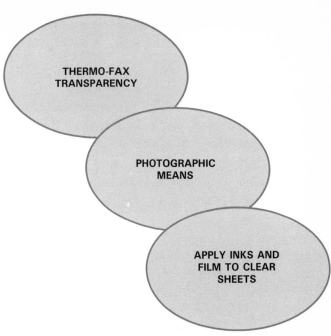
Fig. 14-13. Methods of preparing photographic stencils.

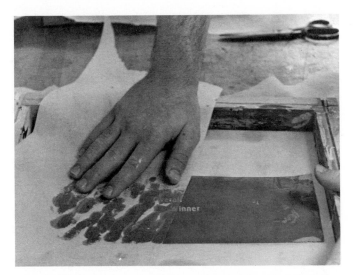
Fig. 14-12. Moisture is best removed with a clean rag or paper towels. The stencil must then air dry before further preparation and printing.

developer in a tray, Fig. 14-15. Agitate (slowly move) the exposed film by rocking the tray back and forth. Again, recommended developing times differ. Refer to manufacturer's and instructor's directions.

Next, hold the film by the corners while lightly spraying the exposed sections of the film. Warm water (about 100°F) will dissolve the image area. The nonprinting area will remain solid.

produce the positive is by photographic means. Use of a process camera helps create very detailed positives. The third method for preparing positives is to apply opaque films and inks to clear plastic sheets.

Once the positive is made, the stencil is produced. This process is called BURNING. The emulsion (dull) side of the film is exposed to light. A platemaker is often used for this process, Fig. 14-14. A high intensity light source (like an overhead projector) may also work for school use. Exposure time will vary depending upon film and light source. The image area that was shielded from the light will wash away during developing.

The exposed film is then developed with chemicals. A special commercially available developer is used for processing the film. Place the

Fig. 14-14. A commercial platemaker. (nuArc Co.)

Adhering film to a framed screen is done much the same way as already described. Water on the film acts as the adhering fluid. Blot excess water up through the film. Dry paper towels or rags work best. A layer of newspaper placed under the film and frame also helps remove water. Newspaper also creates pressure, helping to adhere the film. Use a small roller (called a brayer) to adhere film, Fig. 14-16. Let the screen dry. Remove the clear backing sheet when dry. Peel carefully.

PRINTING PROCEDURES

With the stencil material now placed in a framed screen, the printing procedure can be completed. There are four more important steps. The frame must be assembled, the screen prepared, the print made, and the screen cleaned up.

FRAME PREPARATION

Common sizes of frames can be purchased. These will be placed on a hinged backing board for printing, Fig. 14-17. The screen is usually wedged in place on the frame by ropes. Strong tape or staples are also used. Be sure the screen is

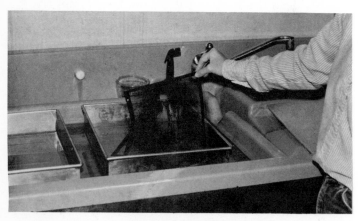

Fig. 14-15. Processing chemicals are placed in trays for developing the film.

Fig. 14-16. A brayer (roller) is useful in adhering the film and forcing water out of the screen. Can you think of other ways to help dry the stencil?

stretched tightly across the frame. If it is not, choose another frame. The screen may need washing before it is attached to the frame.

SCREEN PREPARATION

After adhering the stencil, the screen area outside the stencil will still be open. Ink will pass through this area if it is not blocked. Therefore, use masking tape or heavy paper to block this region, Fig. 14-18. This procedure is called MASKING OFF. Newsprint or cardstock also works well as a mask. For limited runs, only the top side needs masking. Unwanted printing is avoided by blocking out this area. Cleanup is easier too.

Fig. 14-17. Screen frames are attached to hinged backing boards. The frame is lowered into place for printing.

Next, carefully check the image area of your stencil. Small openings may have developed in the film or paper. These must be blocked before printing. A liquid masking fluid is available for this work. In emergency situations, tape will also block the opening. The stencil backing must be removed before applying any masking agent.

The backing sheet of film stencils may be removed at any time. It is wise to leave the sheet in place until you are ready to print. The stencil is protected by leaving this sheet in place. A properly prepared screen is shown in Fig. 14-19.

PRINTING

Printing requires several pieces of equipment. The printing medium (paper, textiles, glass, etc.) should be prepared and in place. The transfer medium must be aligned with registration marks

Fig. 14-18. Non-printing areas should be blocked off. This saves ink and speeds cleanup.

Fig. 14-19. This screen is ready for printing. Can you identify the steps used to prepare this stencil?

on the backboard. Close the frame assembly over this material. A spatula, squeegee, and ink are also needed. The ink selected depends on the stencil material. Water-based inks are preferred for paper stencils. Hand-cut lacquer films permit the use of oil- or water-based inks. Stencils from aqua or photo indirect film require lacquer or oil-based inks. The proper solvent (water or lacquer thinner) should be handy also. This is useful for cleanup purposes.

Stir the ink or paint with a spatula. The color must be well mixed. Apply some ink to the top portion of the screen. The deposit of ink should be placed just above the image area. Do not use too much ink. You can always get more from the can.

A squeegee is used to spread the ink, Fig. 14-20. Squeegees come in various shapes and sizes. The shape of the blade is most important. A square edge is used for printing on paper or other flat surfaces. A rounded squeegee blade is recommended for textile work. The squeegee should also be the proper length. It must be just wide enough to cover the image area. If it is too wide, it will ride on the masking material. If the squeegee is too narrow, two passes will have to made. A squeegee that is too wide or too narrow will result in a poor print.

Next, pull the squeegee firmly across the stencil to force ink through the screen. Again, a single pass is preferred. However, thick ink or a large image area may demand additional passes. Allow the print to dry before handling or placing another color down.

Industrial screen process printing is usually done by machine. Mechanical arms pull the squeegee across the stencil. Special machines allow the printing of round objects. This process of graphic reproduction is used in many areas, Fig. 14-21.

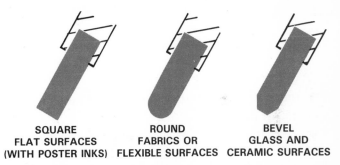

| SQUARE FLAT SURFACES (WITH POSTER INKS) | ROUND FABRICS OR FLEXIBLE SURFACES | BEVEL GLASS AND CERAMIC SURFACES |

Fig. 14-20. A squeegee is used to force ink through the stencil. Different shapes aid the printer in screening a variety of materials.

CLEANUP

After printing, your screen and supplies must be cleaned. A piece of cardstock can clear excess ink from the screen. (Remove any extra ink from the spatula and squeegee too.) Next, peel away the mask and discard. A rag with the proper ink solvent is used to clean the screen, Fig. 14-22. The printing area must be completely clear of ink. Most screens are reusable with adequate cleaning. The squeegee and spatula must be cleaned also.

Finally, the stencil must be taken off the screen. That way the screen can be used later for another job. Place old newspapers under the screen. Pour some stencil solvent (water or lacquer thinner) on the area. Bleach and hot water will remove photo

2 COLOR

Fig. 14-21. This large, industrial screening machine is used in printing two-color jobs. (American Screen Printing Equipment Co.)

Fig. 14-22. Clean off the screening equipment with the correct ink solvent.

indirect stencils. The dissolved stencil material will come off the screen onto the newsprint. Lightly rub solvent over the screen until clean. Check to see that the screen is clear by holding it up to the light. Repeat this procedure if necessary.

SUMMARY

Screen process printing involves forcing ink through a stencil. Screen stencils can be prepared by either hand or photographic means. The best method to use depends on available equipment and print quality desired.

Screens and stencil are fairly easy to prepare. Each stencil preparation method has various steps to follow. After the stencil is made the remainder of the printing process can be done. This includes frame and screen preparation, printing, and cleanup.

KEY WORDS

All the following terms have been used in this chapter. Do you know their meanings?

Adhering, Aqua film, Burning, Emulsion, Film stencil, Gelatin coating, Guides, Lacquer film, Masking off, Paper stencil, Photo indirect stencil, Platemaker, Registration marks, Screen, Spatula, Squeegee, and Stencil.

TEST YOUR KNOWLEDGE — Chapter 14

(Please do not write in this text. Place your answers on a separate sheet.)

1. Screen process printing involves forcing _____ or _____ through a _____ .
2. List several disadvantages of screen process printing.
3. Stencil preparation methods include:
   a. Photo indirect.
   b. Hand-cut paper.
   c. Hand-cut film.
   d. Tusche.
   e. All of the above.
4. What is the difference between a registration mark and a guide?
5. Name the two types of film used in hand-cut film stencils. How is each type adhered to the screen?
6. When preparing a hand-cut film stencil, only the gelatin coating is cut. True or False?
7. What is used to produce photo indirect stencils?
8. Define burning. When is it used?
9. The printing procedure includes:
   a. Cleaning the screen.
   b. Assembling the frame.
   c. Preparing the screen.
   d. Making the print.
   e. All of the above.
10. A _____ is used to stir ink or paint and a _____ is used to spread it.

ACTIVITIES

1. Make your own screen process printing frames. They may be mass produced by the entire class. Use several different methods to attach the screen fabrics to the frames.
2. Design and print a three-color greeting or holiday card using the hand-cut stencil method.
3. Collect samples of different products that use screen process printing in their manufacture.
4. Visit a local screen process printing shop.
5. Bring in your own T-shirts that were made by screen printing. Use them as a basis for discussion on the types of screen printing techniques.
6. Design and print one-color posters for a school spirit activity.
7. Design and print decals using the photo indirect screen process printing method.

15 Lithographic Printing

The information given in this chapter will enable you to:
- ○ *Describe the purposes and uses of lithographic printing.*
- ○ *Prepare masters for lithographic printing.*
- ○ *Produce a variety of products using the lithographic process.*

Lithographic printing is the most widely used reproduction technique in the printing industry. It is commonly known as OFFSET PRINTING or OFFSET LITHOGRAPHY. This method prints (reproduces) images from a flat surface. Lithographic printing is based on the principle that grease and water do not mix.

In the lithographic process, an original design is first prepared. A MASTER PLATE is produced from this design for the printing process. The image area is made up of a greasy substance. The nonimage area is kept clean. This nonimage area will attract water. The greasy, image area will attract ink. The master is first dampened with water. Then ink is applied to the surface. The ink, which only adheres to the image area, produces an image when brought into contact with paper.

Originally, early printers had trouble with this procedure. Their master (usually carved on stone) had to be produced backwards. Therefore, a system of printing the image twice was developed. This permitted right-reading masters, Fig. 15-1. With the offset press the image is transferred to a second surface before printing, Fig. 15-2.

PURPOSES AND USES

Lithographic printing is used primarily when a large quantity of copies are needed. This method of reproduction is fast and fairly inexpensive. Also, detailed photographs can be printed with

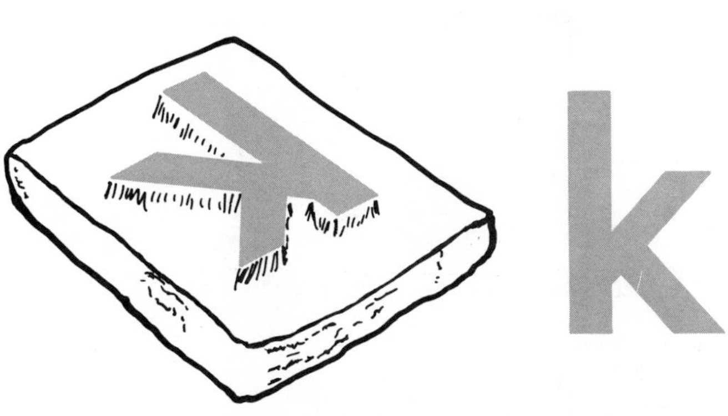

Fig. 15-1. Left. Early stone masters had to be created "wrong-reading" or backwards. Right. By printing the image twice, the master can be designed right-reading.

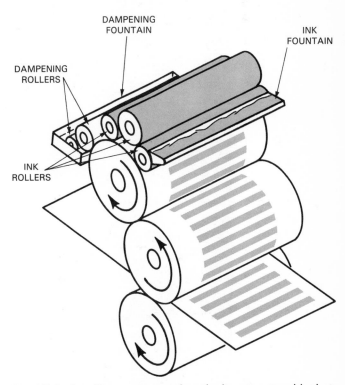

Fig. 15-2. An offset press transfers the image onto a blanket roller before printing on the actual sheets of paper.

excellent results. Jobs requiring various colors are easily reproduced, too.

The lithographic process is used widely in industry. Most industry work includes printing on paper or cardstock. These include books, flyers, envelopes, cards, and reports, Fig. 15-3.

Newspapers are also commonly produced by offset methods. Large rolls of paper are fed into the presses. As it flows through the machines, it is printed, folded, cut, and assembled. Modern printing techniques allow color photographs to be reproduced, too. These huge machines often cost over a million dollars, Fig. 15-4.

The lithographic process is also used in printing posters, magazines, catalogs, and packages. Many of these contain colorful graphics. Only one color is printed at a time. The four most used colors are red (magenta), yellow, blue (cyan), and black. Magenta and cyan are the technical names for two important colors. Magenta is really a red/blue shade and cyan is a blue/green shade. These four colors blend to produce many other colors.

Packaging is designed to attract attention. This is done by using colorful graphics. Many con-

Fig. 15-4. This employee checks the quality of print in the day's newspaper. The machinery in the background is used to print and assemble the papers.
(San Angelo Standard Times)

Fig. 15-3. You probably recognize many of these common items. They were all printed by offset means.

tainers are printed by lithographic means. Large presses often lay down several colors one right after another, Fig. 15-5. This speeds up the printing process.

Besides printing on paper, lithography is used for beverage cans. Sheets of metal are fed through the press. Designs and lettering are transferred onto these flat sheets. Later the metal is cut, rolled, and assembled, Fig. 15-6.

MASTER PLATE PREPARATION

Early lithographic printers used large stones as the surface for printing. Modern offset lithography relies on flexible plates, Fig. 15-7. Three types of plates are generally used in schools and small industries. These are DIRECT-IMAGE, PHOTO-DIRECT, and PRESENSITIZED plates. Direct-image and photo-direct plates are made of paper.

Fig. 15-5. Some larger offset machines fill an entire room. They permit the printing of thousands of images each hour. (Advanced Process Supply Co.)

Fig. 15-6. Many containers are printed with the aid of offset presses. This large roll of aluminum will eventually become cans used for the packaging of food items. (Reynolds Metals Co.)

Fig. 15-7. A lithographic plate.

Presensitized plates are made of metal. The quality and quantity of the printing job determines the best plate.

DIRECT-IMAGE PLATES

Direct-image plates are the easiest to prepare but produce the lowest quality of lithographic prints. This procedure uses paper plates (most are purchased commercially). Various methods are used for creating images and text. Images can be hand drawn with special reproducing pencils and pens. These instruments leave a deposit on the plate that will absorb ink. Typewriters with carbon-based ribbons are also used, Fig. 15-8.

Direct-image plates contain light blue guidelines. These lines identify the image area and aid in centering and spacing. You may also make your own guidelines with a nonreproducing pencil. Be sure to mark the surface lightly. Then trace over the artwork with your reproducing pencil. Again do not press too hard. Mistakes may be erased with soft erasers. Also, keep the surface area clean. Oils and dirt (including that from your fingers) on the plate will cause problems. Handle the plate as little as possible.

The major advantage of direct-image plates is the ease of creating the master. Lettering can be typed or handwritten directly on the plate. Many simple projects (or jobs) can be produced by this process.

PHOTO-DIRECT PLATES

Photo-direct plates are produced quickly but have several limitations. The major drawback is the limited number of copies that can be produced. These plates are designed for runs of under 100 copies.

To prepare photo-direct plates an original print is required. This print can be typewritten copy or artwork. It must be clear and dark (black is preferred). Also, a platemaker is required, Fig. 15-9. The platemaker takes a picture of the original design. An image is projected onto a light sensitive paper plate. Paper plates come in roll or sheet form. Some machines will develop the plate internally. Others require that another processor be used.

After the plate is exposed and developed, it is ready for printing. The plate is loaded on the press by hand. Commercial offset presses with attached platemakers are available. The original is automatically fed into the platemaker and onto the press.

PRESENSITIZED PLATES

The presensitized plate process is more complex than either the direct-image or photo-direct methods. In addition, it requires some expensive equipment. However, a very high quality of reproduction is possible with these plates.

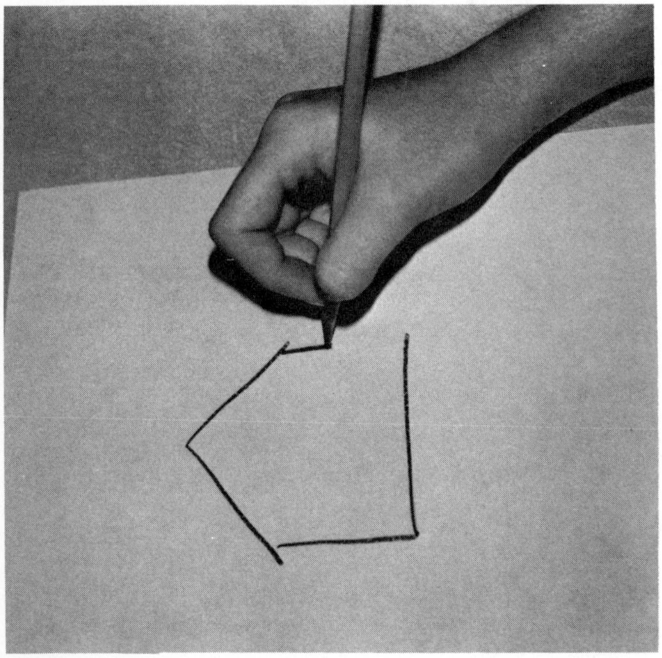

Fig. 15-8. Direct-image plates are inexpensive and simple to use. Artwork and text are easily created with special pens and inks.

Fig. 15-9. This modern platemaker is typical of those found in industry. Do you have a similar model at your school? (nuArc)

Presensitized plates are usually made of aluminum. Many have both sides prepared for printing purposes. This permits images to be developed on either side, Fig. 15-10.

All copy and artwork for presensitized plates must be created with opaque (black) markings on a white background. A line and/or halftone negative is produced from this original. A HALFTONE NEGATIVE is actually a photograph that is converted into a series of dots.

Making line or halftone negatives requires a process camera, darkroom, and developing chemicals. These are shown in Fig. 15-11. Also, orthochromatic film is needed. This provides a high contrast transfer medium. Chemicals required in developing must be mixed and arranged in the order of their use, Fig. 15-12. Process photography uses a developer, stop bath, fixer, and rinse water. Refer to a manufacturer's or instructor's directions for preparing the chemicals.

A process camera is used to produce the negative, Fig. 15-13. The original artwork is placed in a copy board. Film is placed in the camera. The many dials and controls on the camera are used to adjust exposure time, f-stop (lens opening size), and image focus. The percentage of enlargement or reduction must also be set. A proportion scale and exposure table are helpful in establishing the

Fig. 15-12. The chemicals needed to develop film must be prepared before work begins. These containers are arranged in order of developing.

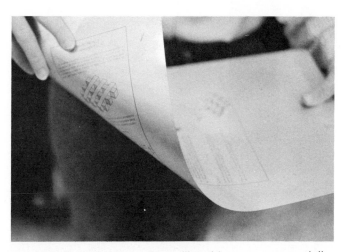

Fig. 15-10. Aluminum plates like this are commercially available for lithographic work. Both sides are presensitized for easy usage.

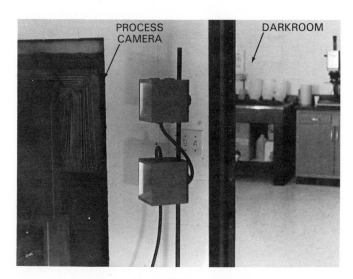

Fig. 15-11. Most photographic work for offset printing requires a darkroom facility. The area shown here has the process camera and darkroom together. This equipment is used to produce the halftone negatives needed for making offset plates.

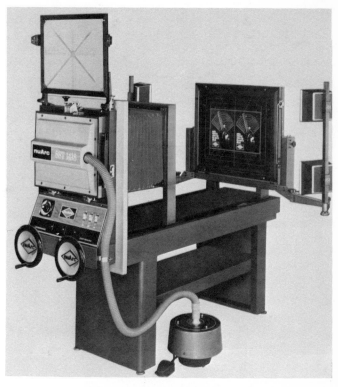

Fig. 15-13. The horizontal process camera has many features. Can you identify the copy board, lights, and main body of the camera? (nuArc)

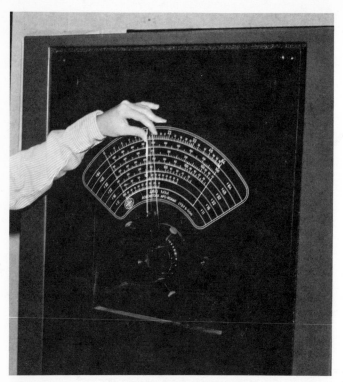

Fig. 15-14. Process cameras have many controls to adjust for different films and light conditions.

Fig. 15-15. A photographic gray scale.

proper settings. The common controls on the process camera are shown in Fig. 15-14.

When ready to photograph the original artwork, it is best to follow certain procedures. The glass on the copy board should be cleaned first. Dust both sides of the glass. The original is then centered under the glass. Place a GRAY SCALE near the copy, Fig. 15-15. A gray scale is a strip of film with a range of grays from clear to totally black. It aids in exposing the film the proper amount of time. Now set the f-stop and exposure time.

The next step is to load the camera with orthochromatic film. Remember that photographic film is light sensitive. This means that film must be handled in a darkroom under safety lighting. Safe lights are normally red. Open the film box under these conditions only. Carefully pick up the film by the edges. Then cut the sheet of film to the desired size. Film is placed on the camera's vacuum board with the light (emulsion) side up. After the film is loaded and centered, close the camera. You are ready to start the exposure.

Remove the exposed film (again handling only the edges). Place the film in the developer tray (emulsion side up). Agitate the tray so the developer covers the entire sheet. An image will slowly appear. Film is usually developed for 2 3/4 minutes or to a solid step 4 on the gray scale. Then place the film in the stop bath for about 10

seconds. The film is next put in the fixer for 2 to 4 minutes. From the fixer, wash the print in the rinse water tray. This clears away the developing chemicals. Leave in this tray for at least 10 minutes.

Following the rinse cycle, the film must be dried. This can be done by hanging it up in a dust-free area, or placing it in a vacuum film dryer. Avoid getting water spots on the negative. The film is now ready to prepare in a flat.

Lithographic plates are produced using this sheet of film assembled in a flat. A FLAT is a ruled (lined) goldenrod sheet, Fig. 15-16. The film is taped to the back side of the goldenrod sheet. A small window (hole) is cut in the sheet. Opaque red tape is used to secure the film. Guidelines on the sheet help register the print. The assembly of a flat is shown in Fig. 15-17.

The presensitized plate may now be exposed. A platemaker is needed along with several developing supplies. These include desensitizer, developer, cotton pads, and gum arabic. Running water is also required. Place the assembled flat on top of the plate (right-reading). The gripper margin at the top should be aligned. Place them on the vacuum table of the platemaker. Be careful so you do not move them out of alignment. Close the glass and

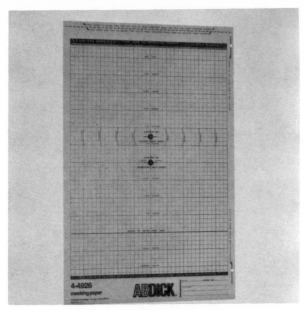

Fig. 15-16. The goldenrod sheet used for producing a plate is called a "flat".

Fig. 15-17. This film is being taped into place on the flat. The goldenrod flat will next be attached to the plate.

turn on the vacuum. Rotate the frame and set the timer. Exposure times vary due to light source and type of plate. The carbon arc light of the platemaker burns an image into the plate. Repeat this procedure to expose the back side, if needed.

Developing the plate is fairly simple. Pour a small amount of desensitizer on the surface and wipe gently over the plate. Use a clean cotton pad and rub lightly. While the plate is still damp from desensitizer, apply the developer. Rub it over the plate lightly. As the developer covers the plate, an image will appear, Fig. 15-18. It will be either red or black depending on the type of plate. Once a bold image appears, wash off the plate. A gentle spray of water is best. Allow it to dry for several minutes.

The plate is now ready for printing. If the plate is to be stored, coat the surface with gum arabic.

This protects the image areas. A cotton pad is useful in wiping the fluid over the plate.

PRINTING PROCESSES

Offset presses all operate using the same systems. Each press has a dampening, inking, printing, and feeding and delivery system, Fig. 15-19. The dampening system provides the water solution needed for this process. Ink is applied to various rollers and cylinders inside the press. The printing system actually transfers the image to the

Fig. 15-18. Chemicals are used to develop the plate after it is exposed to a bright light. An image will appear as these chemicals are rubbed across the plate. (nuArc)

paper. The feed and delivery systems move the paper through the press in registration.

PRESS MAKE-READY

Prior to printing, these systems must be prepared. This is known as press make-ready. The fountain solution is mixed and placed in the reservoir. Adjustment of the inking system may take some time. Several control knobs help adjust the inking rollers, Fig. 15-20. Your instructor will provide a demonstration of these procedures.

Before printing, it is wise to remember key safety suggestions, Fig. 15-21. Printing presses are industrial machines. Caution must be observed in using this equipment.

Now, turn the press on at its slowest speed. Check to see that the inking rollers have an even flow of ink. Small thumb screws at the ink fountain adjust the flow. Also, set the dampening system. Then turn off the press.

Next, load the paper in the feed end of the press. The system is usually adjusted to register standard stock pages. Handles (or electric switches) raise and lower the pile of paper, Fig. 15-22. Again, a variety of procedures exist for each type of press. Consult your instructor for specific details.

PRINTING

To prepare for printing, the plate must be etched with a special solution. This solution also removes any gum arabic that might have been applied. Rub the etching solution over the plate with a cotton pad.

The developed plate is then loaded onto the press. Attach the plate to the master cylinder, Fig. 15-23. Clamp the top in place and crease the plate. Rotate the cylinder until the end may be clamped. Some machines have pins rather than clamps for the plates.

Next, turn on the motor that drives the cylinders. The next step may differ. Some machines specify that the moistening controls be engaged next. Other presses simply require the inking controls be started. When properly set, move the switch to "image." This transfers (offsets) the image from the ink plate to the printing blanket. Hold the switch for several revolutions. You are now ready to print. A switch activates the paper feed mechanism, Fig. 15-24.

Fig. 15-20. Control knobs on the press help adjust the flow of ink and water to the rollers.

RULES FOR PRINTING EQUIPMENT
• Read instruction manual carefully before using any equipment.
• Be sure machine is in good working order.
• Safety guards should be in their proper position.
• Proper clothing is important; loose clothing, jewelry, and long hair are hazardous around machinery.
• Keep hands, arms, and fingers away from moving parts.
• Do not make adjustments while the press is running.
• Follow the directions of your instructor in using the printing press.
• Run the press at a safe speed.
• Store inks, solvents, and cleaning fluids in safe containers.
• Discard rags in metal safety cans with tight lids.
• Ask permission before using any equipment or supplies.

Fig. 15-21. Safety rules to follow when using a printing press.

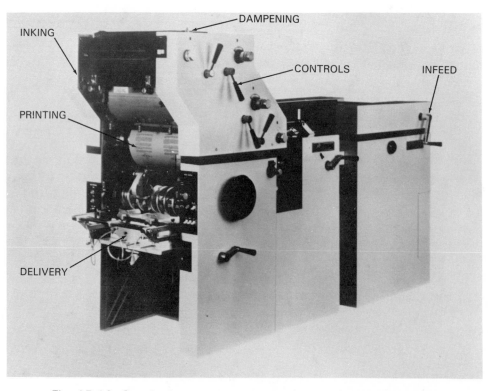

Fig. 15-19. Standard systems of an offset press. (AFT—Davidson)

Fig. 15-22. This employee is loading paper onto the feed system of the press. (AFT—Davidson)

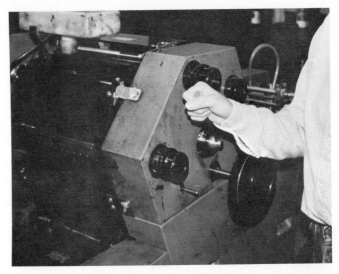

Fig. 15-24. Turn the paper feed control to start the feeding of paper into the press.

Fig. 15-23. The master plate is clamped onto the cylinder of the press as shown.

Fig. 15-25. The counter on the press will keep track of the total number of pages printed.

As paper flows through the press, the image is offset from the blanket to each page. Inspect the printed sheets as they exit from the delivery system. Continue the run if the image is clear and in its proper place (registered). Small adjustments may be required before continuing the run. As you complete the printing work, a counter will total the pages sent through the press, Fig. 15-25.

When finished, turn off the inking and water systems. The control switch is usually turned to "neutral." Use a solvent to clean the impression and blanket cylinders.

PRESS CLEANUP

The press will probably not need further cleanup at this time. However, if you plan to change colors, cleaning is necessary. It is also important if the press will not be used for some time. Use care during cleanup. Do not let fingers or rags get caught in a moving press.

Drain and remove the water fountain first. Remove the ink fountain and clean it. A rag with the proper solvent is best for cleaning purposes. Carefully apply solvent directly to the press rollers as they slowly rotate. Cleaning sheets should also be used, Fig. 15-26. Some rollers are removable for easy cleaning. Wipe the blanket and impression cylinder with solvent on a clean rag. Then reassemble all components.

Finally, clean your hands and dispose of soiled rags. Greasy rags are best kept in safety containers. Discard or store unwanted paper. Also, store all supplies.

SUMMARY

In this chapter, you learned about lithographic printing. It is commonly called offset printing.

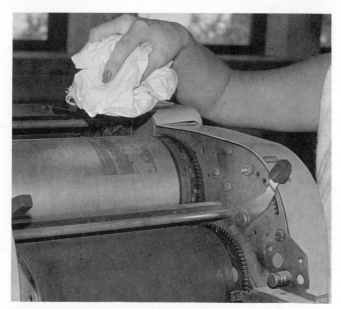
Fig. 15-26. It is important that your offset press be properly cleaned and maintained.

This process is based on the principle that grease and water do not mix. It is used in a variety of products, from posters to beverage cans.

Three types of plates are frequently used in offset work: direct-image, photo-direct, and presensitized plates. Each is prepared in a specific way. Direct-image is the easiest to produce. The presensitized process is the most complex.

The final step in lithography is printing. There are several types of printing presses. The printing procedure, however, is basically the same for all presses. The press must be made ready and then the plate run.

KEY WORDS

All the following words have been used in this chapter. Do you know their meanings?

Copy board, Cyan, Dampening system, Densensitizer, Direct-image plates, Feeding and delivery system, Fixer, Flat, Fountain solution, F-stop, Gray scale, Gum arabic, Halftone negative, Inking system, Lithography, Magenta, Master plate, Offset lithography, Opaque, Photo-direct plates, Presensitized plates, Printing system, Print makeready, Process camera, and Stop bath.

TEST YOUR KNOWLEDGE — Chapter 15

(Please do not write in this text. Place your answers on a separate sheet.)
1. Two other names for lithographic printing are _____ _____ and _____ _____ .

2. Lithographic printing:
   a. Is used when a small number of copies are needed.
   b. Is fast and fairly inexpensive.
   c. Cannot print detailed photographs.
   d. Can use only one color.
3. _____ is a red/blue shade and _____ is a blue/green shade.

MATCHING QUESTIONS: Match the definition in the left-hand column with the correct term in the right-hand column.

4. __ A photograph that is converted to a series of dots.
5. __ Easy to prepare, but of low quality.
6. __ A strip of film used to determine the proper exposure time.
7. __ Plates made of aluminum.
8. __ A ruled goldenrod sheet.
9. __ Plates designed for runs of under 100 copies.

a. Gray scale.
b. Photo-direct plates.
c. Flat.
d. Presensitized plates.
e. Halftone photograph.
f. Direct-image plates.

10. What is the proper order of chemicals for film development?
    a. Rinse water, developer, stop bath, and fixer.
    b. Fixer, developer, stop bath, and rinse water.
    c. Developer, stop bath, fixer, and rinse water.
    d. Stop bath, developer, fixer, and rinse water.
11. Name the four operating systems found in all offset presses.
12. What is the proper order of chemicals for developing a presensitized plates?
    a. Developer, water, and desensitizer.
    b. Water, developer, and desensitizer.
    c. Gum arabic, developer, and water.
    d. Desensitizer, developer, and water.

ACTIVITIES

1. Visit a newspaper printing room to observe lithographic printing in progress.
2. Design a note pad. Print it using the direct-image printing method.
3. Design an advertisement for a school event, and print it using the photo-direct printing method.

4. Use a presensitized metal plate to design your personal stationery. Display a copy of each student's design on a bulletin board.
5. Make a worksheet of steps to follow for operating your school's offset press. Include safety precautions!
6. Compose and print a two-color newspaper using the offset method.
7. Research occupations associated with lithographic printing. Write a short report on the occupation that sounds most interesting to you.

This lithographic plate is in place and ready for use.

16 Continuous Tone Photography

The information given in this chapter will enable you to:
- *Describe the purposes and uses of continuous tone photography.*
- *Demonstrate skill in using a camera and producing photographs.*
- *Use the outlined composition techniques to properly take photographs.*
- *Develop film using photographic chemicals and darkroom equipment.*

Photography is an exciting and interesting area of graphic communication. When you develop and print pictures, it feels as if you are performing magic. Photographers of all ages can produce excellent pictures and slides, Fig. 16-1.

Very simply defined, photography uses light to form images on light-sensitive film. A camera focuses rays of light towards the film. A LATENT (invisible) image is created on the film upon exposure. Developing the film with chemicals produces a NEGATIVE.

To make prints (pictures), light is passed through the negative onto light-sensitive paper. We

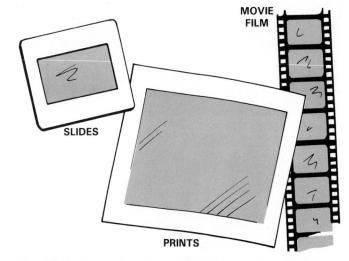

Fig. 16-1. Examples of continuous tone photography are reflected in these common forms of pictures.

call this print a CONTINUOUS TONE PHOTOGRAPH. Continuous tone photographs contain many shades of gray, or variations of color. This projected image is also latent. The paper is later developed to produce the picture. The process of developing pictures is illustrated in Fig. 16-2.

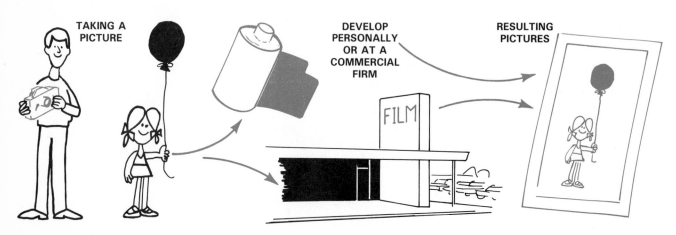

Fig. 16-2. Photographs are taken with camera equipment and developed with chemicals.

PURPOSES AND USES

Photographs record visual images of persons, places, or things. Pictures are truly "worth a thousand words." They transfer information better than either written or spoken words. Pictures also provide a better understanding of magazine or newspaper articles and books. Watching the news on television is much different than listening to the radio. Pictures (video-taped features) bring the news to life. Events seem to unfold before our eyes. This form of communication shapes our emotions and attitudes.

Continuous tone photography has four major uses, Fig. 16-3. These uses include aiding understanding, recording events, decoration, and advertisement. First, photographs help us understand information. For example, this book has photographs to illustrate the topics being presented in the text. Instructional books show "how-to" assemble things through the use of photographs. More complicated or difficult procedures are illustrated for further understanding.

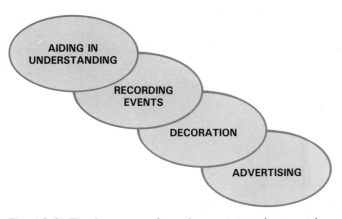

Fig. 16-3. The four uses of continuous tone photography.

Fig. 16-4. The pictures in a school yearbook help us remember school events.

Fig. 16-5. This newspaper advertisement announces a sale to attract more customers to the store.

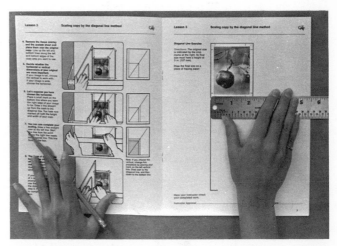

Fig. 16-6. Many corporate brochures have photos to illustrate important points. This booklet will contain several pictures along with text.
(Graphics Arts Technical Foundation)

Photographs are also used to store or record events. Pictures of weddings, vacations, parties, and family gatherings are typical examples. At school, a yearbook is produced to collect memorable scenes, Fig. 16-4.

Photographs are often used to decorate. Pictures of family members can be displayed on walls at home. Friends or special places are photographed, framed, and hung as a reminder. Works of art are placed in museums and galleries.

Finally, photography is used in advertising. Through photographs, a company can show their products and services. Photos of different items often appear in newspaper or catalog advertisements, Fig. 16-5. These pictures help interest a person enough to buy the product or service. Companies project a certain "image" through attractive photographs. Their advertisements remind the public of their service to the community, Fig. 16-6.

CAMERA USE

A camera is a box that controls the transfer of light to film. Complex cameras (35mm, 70mm, etc.) are best at controlling light, resulting in superior photographs. Most cameras consist of two major parts: the box and the lens, Fig. 16-7. The simplest cameras have no adjustable parts on either the box or the lens. Complex units, however, often have many controls. These controls permit clearer and sharper photographs to be taken. The cameras can be adjusted for exposure time. The lens may be focused for improved photographs.

There are several cameras commonly used in commercial and private photography. These include the 110mm (pocket camera), disc, single lens reflex (SLR), twin lens reflex, press, and instant picture cameras. Our explanation of photographic practices will be limited to the 35mm SLR camera, Fig. 16-8.

Before taking pictures, you must know how to use the camera. The instruction manual describes specific procedures. It gives instructions for focusing and loading the camera. In addition, unique features will be pointed out for that particular model. It is important that this manual be reviewed before operating the camera.

Fig. 16-8. 35mm cameras are used most often by both amateur and professional photographers.

Using the proper type of photographic film is important. Film comes in many forms, Fig. 16-9. Common formats include 135mm rolls, 110mm cartridges, and disc cartridges. These may be in either black and white or color, and come in slide or print format. The number of exposures (pictures) on the roll of film is specified. The "speed" of the film is also specified on each package. Film speed is measured in ASA units. The ASA RATING indicates the amount of light needed to expose the film. Low ASA ratings (like 100) are slow films and will produce sharp images. Faster films (like 400) are useful for photographing in limited lighting and for photographing moving objects.

Once the correct film has been selected, the camera can be loaded. Disc and cartridge films are simply inserted into the camera unit. Rolled film demands a slightly different procedure. The film must be threaded into a take-up reel, Fig. 16-10. The leader (film extending out of the canister) is

Fig. 16-7. Parts of a typical camera.

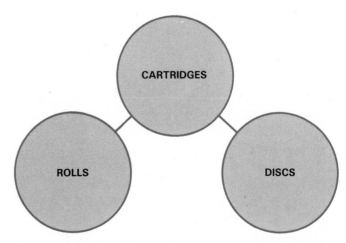

Fig. 16-9. Several formats of commercially available film.

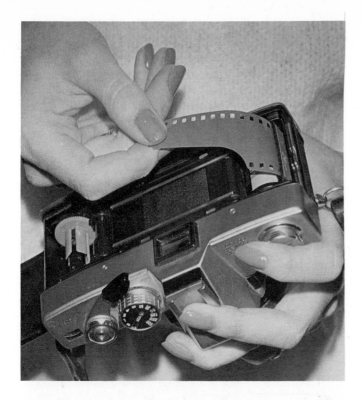

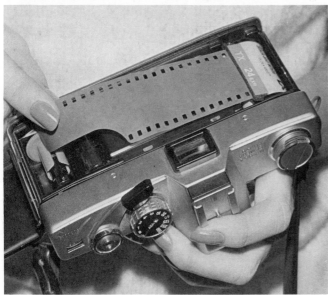

Fig. 16-10. Loading film into a 35mm camera is more difficult than loading film into a disc camera.

Fig. 16-11. The f-stop and shutter controls on a camera. These control the amount of exposure the film will receive.

camera box. (You will recall that photographs are made when this light travels through the lens to the film.) Before pressing the shutter release button, remember to hold the camera steady. Any movement may result in a fuzzy or blurred picture.

The setting of other camera controls is dependent on film and available light. Film speed is set to the ASA rating listed on the container. The shutter opening will determine the amount of light that will strike the film. This opening is called the F-STOP. This is illustrated in Fig. 16-11.

After taking pictures, the film must be developed. Removal of the film from the camera is fairly simple. Rolled film must first be rewound before removing. Turn the rewind knob backwards until the film is completely in the canister. Then open the camera box. Raise the rewind knob to free the film unit. Again, your instructor will have specific instructions for the type of camera you are using.

PHOTO COMPOSITION

Good photographers know a great deal about photo composition. They take their time when composing pictures, Fig. 16-12. PHOTOGRAPHIC COMPOSITION means arranging or putting pictures together beforehand. This is a design stage in the communication process.

Photographs are composed by looking through the viewfinder. Study the scene for the best camera angle and range. Obstructions (distractions in the photograph) should be avoided. These are a form of interference, Fig. 16-13. They can usually be avoided by simply moving the camera to another

attached to a slot in the take-up reel. The film is advanced several times (this may require pushing the shutter release) to be certain of proper loading. The camera is then closed and locked before taking any pictures.

With the film loaded, it is now time to frame and shoot the photograph. A VIEWFINDER is used in selecting scenes. Look through the viewfinder to determine what you are going to photograph. Frame it carefully. When the shutter release button is pushed, light will enter the

Fig. 16-12. These photographers are attempting to "capture the action" on film.

Fig. 16-13. Photographs must be carefully planned to eliminate clutter. This photo was taken too far from the action, making it hard to see important details.

location. This is illustrated in Fig. 16-14. Turning the camera to a vertical position may also help in the composition process.

The principle of balance is critical for composition. Smaller objects or persons should be placed near the front of the view. Avoid exact centering of landscapes or people. This tends to make the photo look posed or unnatural. Proper centering is achieved by using the "thirds principle," Fig.

Fig. 16-15. Divide any scene into "thirds" to identify where the major details should appear. This sign is best when placed above the true center of the print.

POOR

BETTER

Fig. 16-14. Turning the camera often results in a better picture. Notice how distracting trees are cut out by the long, vertical view.

154

16-15. For example, faces should be in the upper one-third for balance, Fig. 16-16.

Remember, good photographs require planning. Visualize what you want to photograph. Compose the scene by looking through the viewfinder. The negative you produce represents the coding phase of communication.

FILM DEVELOPMENT

After the roll of film is exposed, the next step is developing. This procedure is also called PROCESSING. The simplest film to process is black and white negative film. Our discussion will be limited to black and white film processing.

To start film processing, you will need the following: an exposed roll of film, a developing tank, a film reel, a darkroom or film change bag, and developing chemicals. These materials are shown in Fig. 16-17.

Perhaps the most difficult part of film development is loading the film reel. This is because exposed film must be loaded in total darkness. A darkroom or a film change bag are normally used for this procedure. Naturally, loading the film will require practice. Watch carefully as your instructor demonstrates the procedure. Then, when it is your turn to practice, try loading the reel while

Fig. 16-17. Photographic materials and supplies.

keeping your eyes closed. After you master this step, practice loading the reel in a darkroom or film change bag.

Once you can load the "practice" roll of film on the reel, it is time to try with a roll of "real" film. For this step you will need the film, the film reel, and the developing tank. These should all be placed either in the darkroom or the film change bag. Open the canister. Load the film onto the reel carefully, Fig. 16-18. It is sometimes easier to thread the film if the edges of the film leader are

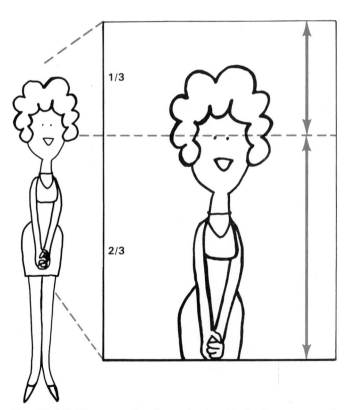

Fig. 16-16. Photographs of people should also be composed with the thirds principle in mind. Head and facial features should appear in the upper sections of the picture.

Fig. 16-18. Film must be loaded onto the reel as shown. Be sure you do this in the dark!

Fig. 16-19. Keep your developing chemicals in marked containers at the proper temperature.

Fig. 16-20. After developing, film is washed while still in the developing tank.

cut in a rounded shape. Now, place the loaded reel into the developing tank. Make certain the lid of the tank is on tightly!

To develop the film, you will need three chemicals: developer, stop bath, and fixer. Handle all photographic chemicals with caution. They can irritate your skin and eyes, and can ruin your clothing on contact.

It is very important to have the chemicals at 68°F (20°C) when developing. Placing the chemical containers in running water will control the temperature. Water temperature is easily measured with a thermometer, Fig. 16-19.

Working near a sink, pour the developer into the tank. Keep track of the time it spends in the developer. Proper development time varies for each processing session. After the chemicals are added to the tank, agitate the container slowly. This assures that developer covers the entire surface of the film.

At the end of the processing time, pour out the developer. Immediately add stop bath to the tank. Leave this in the tank for about 1-2 minutes. Then pour the stop bath back into its container for reuse. Next pour fixer into the developing tank. Agitate the container for 7-10 minutes. The fixer will remove unexposed chemicals from the film. Again, pour this back into its container for reuse. The film can now be exposed to light.

At this point water must be poured into the tank for at least 20 minutes. This washes the chemicals off the processed film, Fig. 16-20. A wetting agent should be added to the water for the last few minutes. This prevents water spots from forming on the film.

Remove the film from the developing tank and reel. It should be hung up to dry, Fig. 16-21. A clip is used to hang the film while drying. At-

Fig. 16-21. Hang up film to dry. Be sure clips are attached to both ends to keep the film from curling up and sticking together.

taching a clip to the bottom will prevent the film from curling. When the film has dried, cut it into sections. Avoid cutting each negative apart. Rather, cut them into groups of two or more for easier handling.

PRINT MAKING

Making a photographic print is a two-step process. First, the negative must be enlarged. Next, it must be developed into a photograph. The entire procedure is referred to as PRINTING.

Making prints requires an enlarger, developing chemicals, and photographic paper for printing. A CONTACT PRINTING FRAME is helpful for

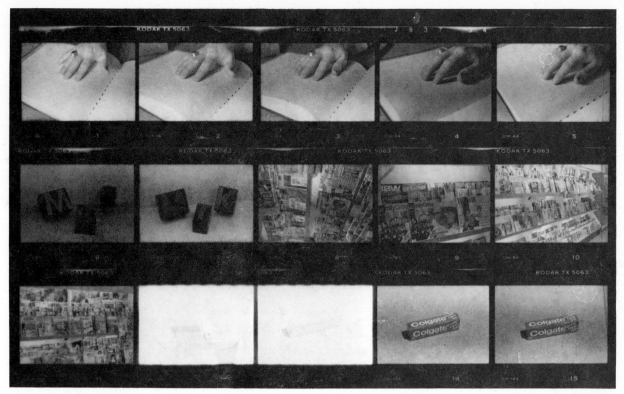

Fig. 16-22. A sample contact sheet allows the opportunity to select the best shot, without making prints of every negative.

inspecting your negatives. This frame allows you to make a one sheet print of all the negatives from a single roll of film, Fig. 16-22. It is likely that you will only want to make full-size prints of your best shots. Contact sheets allow the opportunity to select the best shots.

Assemble the developing chemicals before producing the contact print. The developing and fixing of paper prints requires different chemicals than those used for film negatives. Be sure to prepare the correct chemicals. Place about one inch of each chemical in trays arranged in order (developer, stop bath, and fixer). A bath of circulating water should follow the fixer tray.

CONTACT SHEETS

You are now ready to make a contact sheet. In safe light (red or yellow), remove a sheet of print paper from the box. Remember to close the film box to avoid exposing the remaining sheets. Place the paper, shiny side up, on the enlarger stand. Position the negatives on top of the print paper. The emulsion side of the negatives should contact the paper. Place a piece of glass over the negatives. This holds the paper and negatives in position.

Turn on the light of the enlarger. A timer on the light switch controls the exposure time. Most prints are exposed for about 10 seconds on an f/11

setting. Recommended exposure times and f-stop settings will vary. Consult your instructor for what is best for your negative.

When the light goes off, place the paper into the developer tray. Remember to handle only the corners of the paper. Allow the developer to drain off the sheet before placing it in the stop bath. Developing time usually takes about one minute. Next place the sheet in the stop bath for five seconds. Let it drain again, before putting it in the fixer for two minutes. Finally, put the print in the wash water for a minimum of five minutes.

After washing, examine the sheet for the selection of the best photographs. You may want to cut the contact sheet into strips. Tape these strips to your negative cover for print identification.

ENLARGEMENTS

The process for making enlargements (actual photographs) is similar to producing contact prints. However, the enlarger is used to control the size. It is possible to produce prints of various sizes with an enlarger, Fig. 16-23. You will need to adjust the lens and assembly (head) for each shot.

The negative is loaded into the enlarger head. Remember to place the negative emulsion side down in its carrier, Fig. 16-24. Next, position the

Fig. 16-23. Photographic enlargers allow print size to be increased.

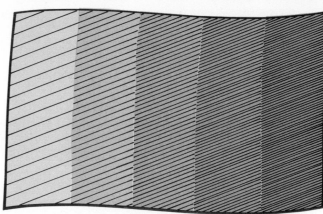

5 SEC.   10 SEC.   15 SEC.   20 SEC.   25 SEC.

Fig. 16-25. The best developing time can be determined by making a test sheet. By slowly uncovering a portion of the paper every five seconds, different time intervals can be compared.

enlarger easel on the stand. Turn the light on to check alignment. The entire surface of the easel should be illuminated. Focus the lens so the photograph appears clear and sharp.

After these adjustments, a TEST SHEET is usually produced to determine exposure time. A sheet of cardboard is needed for this process. Under safe lights, remove a sheet of print paper from its package. Place the paper on the easel shiny side up. Cover all but a fifth of the sheet with your cardboard. Set the timer for five seconds and turn on the enlarger. The light will shut off after five seconds. Move the cardboard so that another fifth of the sheet is showing, Fig. 16-25. Expose the paper again. Repeat this process until all five exposures are made. You will have exposed the whole sheet of paper. The last section will have been exposed for only five seconds, while the first section was exposed for a total of 25 seconds. See Fig. 16-26.

Fig. 16-24. The photographic negative is loaded into a carrier, and then placed in the enlarger.

Fig. 16-26. This test sheet has been exposed from 5 seconds (far left) to 25 seconds (far right).

Repeat the developing process explained earlier. Remember to handle the sheet by the corners only.

Inspect the test strip. Select the time that provided the best contrast (ranges of blacks, whites, and grays). Set the timer for that value. Place a new sheet of paper on the easel. Expose the print for the proper time, then develop.

Your completed print should then be dried. Excess water is best removed with a squeegee, Fig. 16-27. Prints can be air dried, or placed in a drying machine.

As your skill in developing increases, you will become familiar with other advanced techniques. One technique is called CROPPING, Fig. 16-28. Cropping is a common way of removing unwanted sections of a negative from the print. It is done by changing the alignment of the paper and easel.

SUMMARY

Photographs transfer information better than either written or spoken words. The four major uses of photography are to aid understanding, to

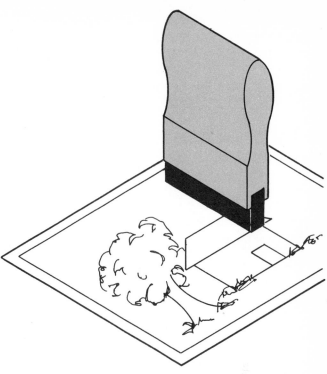

Fig. 16-27. Drying a print with a squeegee.

Fig. 16-28. This photograph will be cropped to leave out clutter and interference.

record events, for decoration, and for advertisement. Using a camera, processing and printing film takes practice. As you practice, following the guidelines in the chapter, you will gain new skills. Your understanding of the principles of photography will grow steadily.

KEY WORDS

All the following words have been used in this chapter. Do you know their meanings?

ASA rating, Composition, Contact printing frame, Contact sheet, Continuous tone photograph, Cropping, Darkroom, Developing tank, Enlarger, Film change bag, Film reel, F-stop, Latent, Leader, Negative, Obstructions, Print, Printing, Processing, SLR camera, Test sheet, and Viewfinder.

TEST YOUR KNOWLEDGE — Chapter 16

(Please do not write in this text. Place your answers on a separate sheet.)

1. Photography uses _____ to form _____ on light-sensitive paper.
2. What is an invisible image called?
3. List the four major uses of photography. Give an example of each.
4. Film speed is measured in:
   a. SLR units.
   b. ASE units.
   c. ASA units.
   d. None of the above.
5. Photographic composition does not include:
   a. The fifths principle.
   b. Balance.
   c. Planning.
   d. Avoiding obstructions.
6. Film development is also known as _____.
7. Outline the steps in the development of film negatives.

MATCHING QUESTIONS: Match the definition in the left-hand column with the correct term in the right-hand column.

8. __ Used to determine correct exposure time.
9. __ A two-step process for making photographs.
10. __ Process of removing unwanted sections from a photograph.
11. __ Used to make different size prints.
12. __ A one sheet print of all the negatives from a single roll of film.

   a. Enlarger.
   b. Contact sheet.
   c. Cropping.
   d. Printing.
   e. Test sheet.

ACTIVITIES

1. Visit a camera/photography shop. Arrange for a salesclerk to show and explain the various equipment and supplies available to the beginning photographer. Take notes and secure brochures. Write a short paper on the camera and equipment you would prefer to use, and why you would prefer to use it.
2. Bring photographs from home. Use these to review composition techniques. Make note of the poor techniques as well as the good techniques.
3. Invite a professional photographer to speak to your class about his or her experiences. Ask your guest to bring sample photographs, as well as the equipment and supplies used. Prepare questions.
4. Choose two or three of your best prints to present to the class. Explain the composition techniques you used. Why do you like these photographs? What makes them good photographs?
5. Give a brief oral report on a famous photographer. Present your report to class, along with several examples of the photographer's work.

17 | Specialty Printing Methods

The information given in this chapter will enable you to:
O *Describe the purposes and uses of specialty printing processes.*
O *Demonstrate skill in using various specialty printing methods.*

Specialty printing methods help communicate and produce our endless flow of information and products. Several of these techniques are very familiar. For example, photocopying is a useful printing process, Fig. 17-1. Ditto and mimeograph methods are common for office and school work. Uses for letterpress techniques are still found in industry. And, of course, typewriters have been around for many years, Fig. 17-2.

SPECIALTY PRINTING IN THE OFFICE

Several copying or duplicating methods are used primarily in office situations. These processes are utilized for several reasons. One reason is because of the limited number of copies required. Low cost and ease of use are other important reasons.

DITTO

The ditto process requires the use of a ditto machine and master, Fig. 17-3. The masters are purchased commercially. They may be direct image or heat transfer. With direct image, the original design is typed or drawn directly on the master sheet. For example, if the design includes written copy, the master sheet is placed directly

Fig. 17-1. Almost every school and business has a photocopy machine. They permit the copying of materials quickly and easily.

Fig. 17-2. The typewriter was invented in the 1800s. Today, much office work is completed with typewriters.

Fig. 17-3. Many school and office handouts are run on ditto machines.

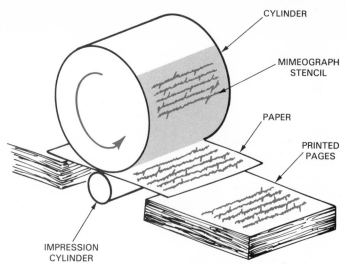

Fig. 17-4. Ink is forced through mimeograph masters. Does this procedure remind you of another printing technique?

on the typewriter and typed on. These masters are available in a variety of colors. Purple is the most common.

Heat transfer masters are prepared differently. The original design is created on standard paper, then photocopied. The photocopy is sandwiched with the master and processed with a Thermo-fax machine. Heat from the light in this machine burns (transfers) the image onto the master. This process is useful for originals which contain artwork.

After preparing the ditto master, it is ready for printing. Attach the master to the cylinder of the machine. Only the cover sheet is mounted on the cylinder (upside down). As the master is rotated, ditto fluid is wiped on it. This softens the image. The wet image is transferred to paper as it contacts the master. The image is partly removed with each copy produced.

MIMEOGRAPH

The mimeograph process is another common office copying practice. It utilizes a stencil much like the screen process printing method, Fig. 17-4. Mimeograph stencils are prepared by direct contact or electrostatic means.

The direct contact method is the simplest of the two methods. An image is typed or drawn directly on the master. This will cut right into the stencil material. In the electrostatic method, a scanner reproduces the original onto a black stencil. Artwork is easily copied by this procedure, Fig. 17-5.

After the stencil is completed, it is positioned for printing. The master is fastened to the

Fig. 17-5. Scanners permit the copying of any combination of text and artwork.

mimeograph cylinder, Fig. 17-6. Ink is forced through the stencil to the paper. A uniform, bold layer of ink is applied to each page.

PRINTER/PLOTTER

The use of computers has led to many types of specialty printing. Modern printers and plotters provide outstanding graphic displays, Fig. 17-7. Many computers act as phototypesetters. These produce camera ready copy in various lettering styles. Dot matrix printers form different typefaces with tiny dots, Fig. 17-8. Computers also direct ink jet printers. This is a modern type of dot matrix printing.

Small business and home computers are often hooked to printers. Many of these printers operate like a typewriter. They create images by forcing a

Fig. 17-6. The mimeograph master is loaded onto the machine's cylinder.

3,.5987,.6179,.6
758,.7734,.7881,
8849,.8944,.9032
9554,.9599,.9641
9861,.9878,.9893
965,.997,.9974,.
993,.9994,.9995,

Fig. 17-8. Dot matrix printers form characters by using small dots.

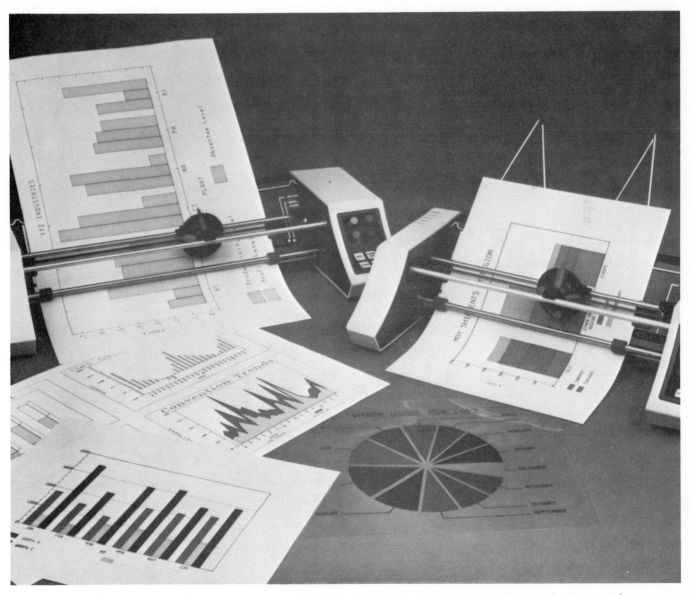

Fig. 17-7. Printers and plotters print words and pictures in several colors. (Houston Instruments)

ribbon against sheets of paper. These are often called DAISY WHEEL PRINTERS, Fig. 17-9. The name comes from the odd-shaped wheel that rotates into position for printing. Letter quality typing is achieved with these printers.

DOT MATRIX PRINTERS are also associated with computers and related machines. These devices form letters and symbols with tiny dots. A set of pins mechanically drive a ribbon into the paper. The quality of image varies upon demand. Cash register slips reflect a low quality. High quality is available from better machines. Many dot matrix printers rival typewriters in quality of image produced.

Any letter or number can be formed from a matrix of dots. Normal arrangements include patterns that are 5 x 7, 7 x 9, or 9 x 9, Fig. 17-10. Different sizes and styles of lettering can be produced by this method of printing. Large and bold titles can be printed along with standard text.

Computers often guide INK JET PRINTERS. Information stored in the computer is transferred character by character onto paper as droplets of ink. The ink is forced through nozzles to form letters and symbols, Fig. 17-11. Again, this is a matrix form of printing. Current uses for ink jet printing include mailing labels and computer printouts.

Fig. 17-9. A daisy wheel printer produces high quality images. Letters and symbols are formed by forcing a ribbon onto the paper.

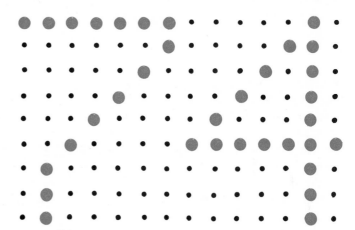

Fig. 17-10. Arrangements of dot patterns used to form characters in dot matrix printing systems.

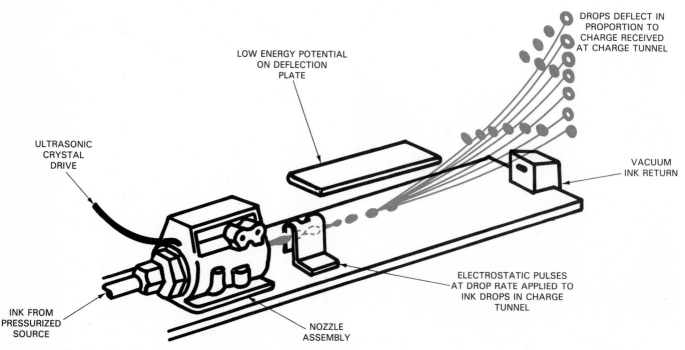

Fig. 17-11. Ink jet printing involves shooting ink onto paper. Each tiny droplet of ink is guided to the page by computer controls. (Marsh Stencil Co.)

164

The ink jet printer is unique in one other respect. Unlike standard typewriters (which force inked ribbons into paper), no parts actually strike the page. Liquid or powdered ink is sprayed on the paper. Then heat cures the inked image as it passes through the machine. These devices are classified as NONIMPACT PRINTERS. Electrostatic copiers function in a similar fashion, Fig. 17-12.

PREPARATION OF PRINTED MATERIALS

Several other specialty printing methods have important applications in the office. These methods are used to develop printed material. Typewriters and Kroy machines help create text for various communication materials. Both machines permit the printing of various lettering styles. This is a large improvement over earlier typewriters, with their single style of letters.

Some modern typewriting equipment is computerized. Data is generated and sent directly to a printing head. In addition, special graphics are achieved through coded instructions. Copies are produced on plates or sheets inserted directly in the machine. This technique is useful in generating forms and other office material. Ease of use and speed are two attractive features of this process.

The use of Kroy machines has increased steadily, Fig. 17-13. These machines allow letters and symbols to be stamped on a clear, gummed tape. Wheels containing various styles of type produce many variations. Graphic designers use Kroy lettering for titles and headings of printed materials. A sample pasteup is shown in Fig. 17-14.

Fig. 17-13. Kroy printing machines produce high quality letters in many typefaces. (Kroy, Inc.)

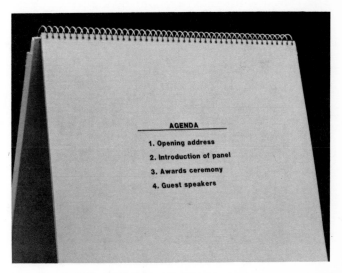

Fig. 17-14. This outline for a business meeting was created with Kroy type before printing. (Kroy, Inc.)

SPECIALTY PRINTING IN INDUSTRY

Many graphic procedures aid in the production of common industrial goods. Clothing, money, and color catalogs are typical industrial products. Each is prepared by a special type of printing process.

RELIEF

You will recall that relief printing uses raised letters to print images. FOUNDRY TYPE is a form of relief printing. It was developed by Johann Gutenberg. Hand or machine setting of foundry type remains popular for a variety of graphic processes. This system of movable type is still useful in the production of rubber stamps,

Fig. 17-12. Electrostatic printers take a picture of the work to be copied. Then the image is transferred to a sheet of paper by electronic means.

wedding announcements, and business cards.

Hand composition of type is generally used for making rubber stamps. Brass foundry type is used because heat is required in the process. Type is stored in CALIFORNIA JOB CASES, Fig. 17-15. Letters are not arranged in alphabetical order. Their location is determined by frequency of use. Commonly used letters occupy larger boxes in the center of the case.

Foundry type for each job is pulled from the case. Type is placed onto a COMPOSING STICK, Fig. 17-16. The nick (or grooved side) is up during composition. Lines of type are divided by SLUGS or LEADING, Fig. 17-17. Spacing within the line is developed with the use of QUADS or SPACERS. The fully prepared type is ready for proofing (checking for errors).

Foundry type is used to make rubber stamps. The type is put together, and heated at approximately 300°F (150°C) for several minutes. The hot type is then forced into a plastic material that is called BAKELITE®. The heat from the type will set the Bakelite® permanently into a mold. This mold is used to form the rubber stamp.

Next, matrix material (rubber) is heated and pressed into the mold, Fig. 17-18. In this instance, the matrix material is a strip of rubber. This vulcanizes, resulting in the surface portion of a rubber stamp. The piece of rubber is cemented to handle stock, Fig. 1⁻ 19. A rubber stamp is an excellent example of the relief printing method.

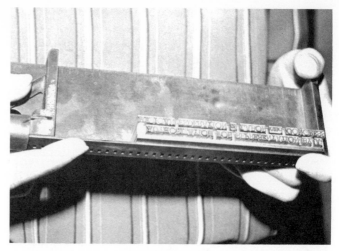

Fig. 17-16. Type is loaded onto the composing stick as shown.

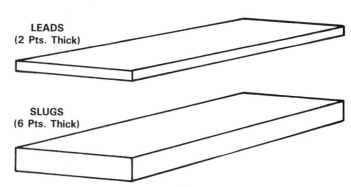

LEADS (2 Pts. Thick)

SLUGS (6 Pts. Thick)

Fig. 17-17. Spacing in hand-set type is achieved with the aid of slugs and leads.

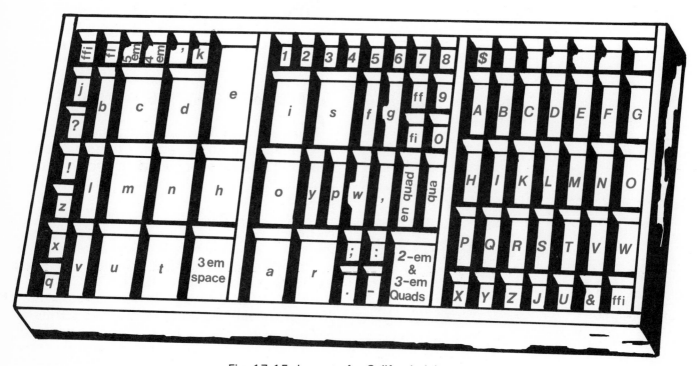

Fig. 17-15. Layout of a California job case.

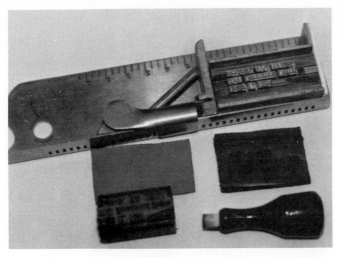

Fig. 17-18. Matrix materials are used in the production of rubber stamps. A strip of rubber is softened by heat and then shaped into the final stamp form.

Fig. 17-19. Rubber stamps are useful for printing quickly and in different colors.

HEAT TRANSFER

Heat transfer is a popular printing method in the textile industry. Many T-shirts are printed using heat transfer, Fig. 17-20. Masters are usually prepared by lithographic or screen process printing. An image is printed onto a backing material. The image is then transferred by heating in a dry mount press or with an iron. There is one important disadvantage to heat transfer printing. They generally do not wear as long as designs screened directly onto the transfer medium.

TEAM NAMES
SCHOOL
CLUB
YOUR NAME
SPONSORS
NUMERALS
SLOGANS
SPECIAL DATES

Fig. 17-20. Decorative T-shirts are often produced with heat transfer designs.

INTAGLIO

Printing from a sunken surface is called intaglio printing. In industry, we call this process gravure printing. Gravure plates contain image areas below the surface, Fig. 17-21. Ink caught in the depressions is transferred to various mediums. Either curved plates or etched cylinders are used in gravure printing.

The gravure process offers many advantages to industry. Modern plates usually permit long runs with consistant tonal quality. High speed presses can print in black and white or color. Artwork and photographs reproduce fairly well also. These characteristics help make gravure printing attractive for many items, Fig. 17-22. Most Sunday newspaper supplements are printed by this method. Currency (money) and color catalogs are also produced by the gravure process. Postage stamps and wallpaper are other examples.

Gravure processes do have several disadvantages. One is the expense of correcting mistakes. Entirely new plates or cylinders must be completed if errors are present. Also, gravure presses are

INK COLLECTING IN GROOVES
OF THE PLATE

PLATE

Fig. 17-21. Images in a gravure plate are below the printing surface.

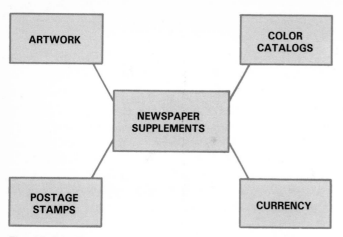

Fig. 17-22. The gravure process is a popular form of printing in industry. Can you think of other products made by this method?

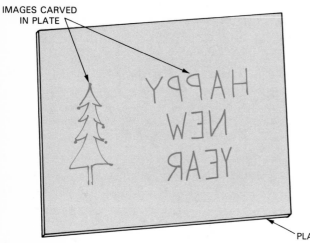

Fig. 17-23. This sheet has been prepared by scratching (etching) lines into the surface. Ink will collect in the grooves.

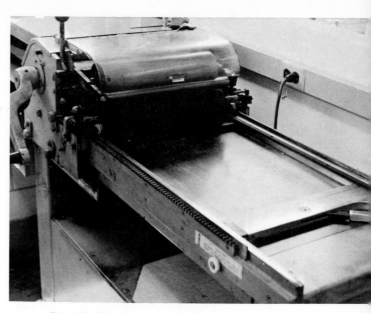

Fig. 17-24. A small industrial proof press.

usually quite expensive. Their cost frequently makes their use impossible in schools.

A form of intaglio printing that is inexpensive enough to use in schools is called DRY POINT ETCHING. Lines are scratched in clear plastic sheets. Usually the image is traced from an original, Fig. 17-23. The surface is then covered with ink. The scratched lines fill with ink. Any ink in nonimage areas is removed. A paper towel will clean the surface. Wipe perpendicular to engraved lines so ink remains in the grooves. If necessary, slightly moistened paper will pull ink from the etched areas. Printing is best done on a small proof press, Fig. 17-24.

SUMMARY

There are many types of specialty printing processes. Basically, they are divided into two categories: those used in offices and those used in industry.

Printing processes used in offices are best suited to this situation for several reasons. The number of copies needed is usually limited. The cost of using these processes is relatively low. And these processes are quick and easy to use. Ditto, mimeograph, and printing machines are printing processes often used in offices. Typewriters and Kroy machines also have important applications in offices.

Specialty process printing in industry includes relief, heat transfer, and intaglio processes. These are used to make items such as rubber stamps, business cards, and T-shirt designs.

KEY WORDS

All the following words have been used in this chapter. Do you know their meanings?

Bakelite®, California job case, Composing stick, Daisy wheel, Direct contact printing, Ditto, Dot matrix, Dry point etching, Electrostatic printing, Foundry type, Gravure, Heat transfer, Ink jet, Intaglio, Kroy machines, Leading, Matrix, Mimeograph, Nonimpact printers, Photocopying, Proofing, Proof press, Quads, Relief printing, Rubber stamp, Slug, Spacer, and Typewriter.

TEST YOUR KNOWLEDGE — Chapter 17

(Please do not write in this text. Place your answers on a separate sheet.)
1. Ditto masters may be _____ _____ or _____ _____ .

2. Mimeograph stencils are prepared by (check all that apply):
   a. Electrostatic means.
   b. Dot matrix printers.
   c. Relief printing.
   d. Direct contact method.
   e. None of the above.
3. _____ _____ printers are classified as _____ printers because machine parts never strike the paper.

MATCHING QUESTIONS: Match the definition in the left-hand column with the correct term in the right-hand column.

4. __ Plastic material used to make a matrix.
5. __ Form of printing developed by Gutenberg.
6. __ Type is placed on this for composition.
7. __ Place where foundry type is stored.
8. __ Mold used to make the surface of a rubber stamp.

   a. California job case.
   b. Composing stick.
   c. Matrix.
   d. Movable foundry type.
   e. Bakelite®.

9. Identify two disadvantages of gravure printing.

ACTIVITIES

1. Choose a specialty printing process to research. Write a paper about the process, based on your research. Include information on the invention, inventor, and history of the process. List several items for which it is currently used. Present the report to class. Secure examples of items made by this process.
2. Visit a shop that manufactures heat transfer designed clothing. Arrange to tour the entire business, from the production of the stencils to the heat transfer of the design.
3. Collect samples of computer generated printing. Identify each sample. Which has the best quality? Secure samples of various typewriter lettering styles. Which of these has the best quality?
4. Make a rubber stamp with your name and address as the design.
5. Design a two color flyer for an upcoming school event. Print it using ditto masters and a machine.
6. Compose a one page class newspaper. Use a Kroy machine to make the masthead. Use typewriters to compose the stories. Use different types of printers (different letter styles) to make the headlines.

18 Introduction To Electronic Communication

The information given in this chapter will enable you to:

○ *Identify the major areas of electronic communication systems.*
○ *Learn of the historical developments of electronic communication.*
○ *Discuss current developments in electronic communication.*

Electronic communication involves the use of electrical energy to transmit information between individuals or systems. This electrical energy can take many forms. Radio messages are transmitted through signals, Fig. 18-1. Light waves are used to transmit telephone messages along fiber optic cables. Neither of these messages could be transmitted without the use of electricity.

We have not always been so knowledgeable about electronic communication. In the past, nearly all communicating was done through printed materials: letters, newspapers, etc. But as scientific knowledge increased, a second communication system, electronic communication, began to develop. This development was, and continues to be, full of exciting discoveries.

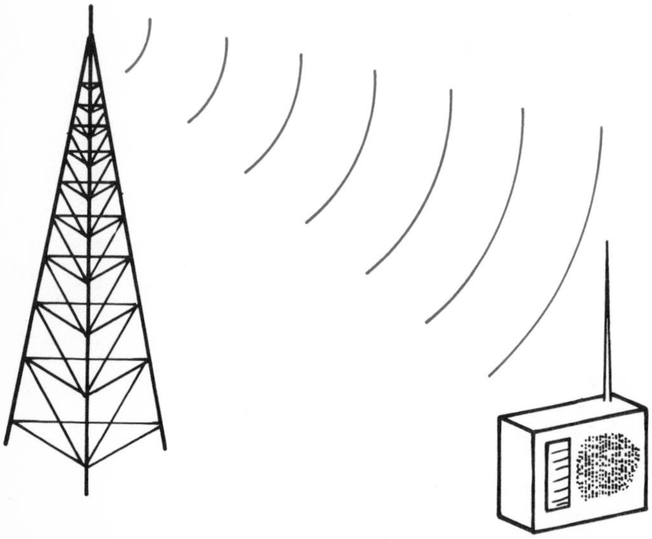

HISTORY OF ELECTRONIC COMMUNICATION

You learned briefly about several of the more important communication devices in Chapter 2. In this chapter we will explore more fully the history and current developments of electronic communication.

EARLY TRANSMISSION TECHNIQUES

Among the most important developments in history are early experiments with the telegraph and telephone. These inventions allowed humans to send messages between remote locations. Of course, simple methods of communicating over long distances existed before the invention of the telephone and telegraph. The Indians of the Great Plains used smoke signals, Fig. 18-2. Church bells and flags were common in most towns. At sea, cannons and lights were used to signal to other ships. Postal systems allowed the exchange of letters and packages in the Colonies during the 1700s. And the famous Pony Express system carried messages across the continent for several years.

However, experiments with electricity in the 1800s provided new potential for communicating between two points. No longer were humans limited to the range of their voices or line of sight, Fig. 18-3. Messages could now be sent along wires in the form of electrical current. This is still the basis for modern telephone technology, Fig. 18-4. Spoken words are changed to electrical pulses for

Fig. 18-1. Radio receivers actually "catch" transmission signals (waves) as they pass antennas.

Fig. 18-2. Indians used smoke signals to communicate over long distances before the invention of electronic communication systems.

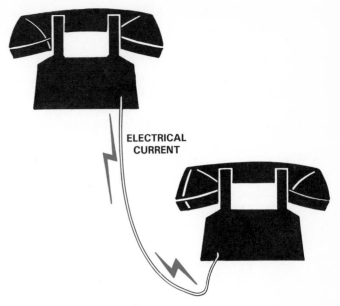

Fig. 18-4. Sounds travel along telephone lines in the form of electrical current.

Fig. 18-3. The human voice can only be heard over short distances. With the aid of a telephone, you may talk to others anywhere on earth.

transmission. These transmissions are received at the opposite end and changed back into a recognizable tone.

EXPERIMENTS WITH RADIO WAVES

From 1886 to 1889, Heinrich Hertz conducted many laboratory experiments with electrical sparks. His work led to the discovery of ELECTROMAGNETIC ENERGY, including RADIO WAVES. Guglielmo Marconi used Hertz's ideas in developing a simple radio signal generator. This invention became known as the "wireless." It was given this name because sounds were transmitted without the use of wires. The wireless was tested successfully in 1895. Marconi's first broadcast across the Atlantic Ocean took place in December, 1901. The world suddenly had a new means of long

distance communication.

Naturally, many problems plagued these early transmissions. The signals were still in Morse code so only a trained operator could translate messages. Eventually, methods for changing voices to electromagnetic energy enabled all people to communicate by radio. Another challenge included designing an antenna that could pick up the faint signals. The almost football-sized antennas used by Marconi were much too large. Later technical developments in radio receivers reduced the size considerably.

Radio station broadcasts have been operating in the United States since 1920. These broadcasts have five distinctive features that identify them as radio broadcasts. They are wireless, transmit to the public, provide a continuous program, and are licensed by the government. In addition, the signal is said to be transmitted by TELEPHONY. This means sounds (for example, music or talking) are not coded but are easily recognized by the listener.

The FM system of radio transmission was perfected in the 1930s. This provided listeners with a signal that remained unaffected by weather and related static. Stereo signals also became possible with the invention of FM radio. However, the most important development to affect radio was the invention of the transistor.

TRANSISTOR TECHNOLOGY

In the early years of wireless transmissions, transmissions were made possible through the use of the vacuum tube. Vacuum tubes act like electronic switches and amplifiers. They produce the electronic frequencies necessary to carry the human voice. However, vacuum tubes have many disadvantages. They are often large and noisy. They operate at high temperatures. It takes time to heat them to this temperature. And they use a great deal of power while in operation.

In 1948, researchers at Bell Laboratories developed a new device that soon replaced the vacuum tube. Experiments by John Bardeen and William Shockley led to the invention of the TRANSISTOR, Fig. 18-5. Transistors are powerful amplifiers. They also perform the same switching function as vacuum tubes. Modern transistors are made of germanium or silicon. Most are very small and inexpensive.

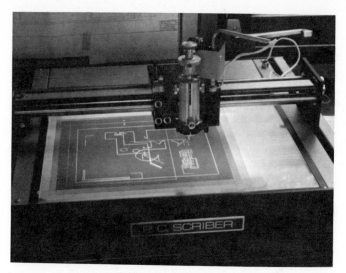

Fig. 18-6. This printed circuit board is being designed by machine. It will be reduced to required size during the manufacturing process. (AT & T)

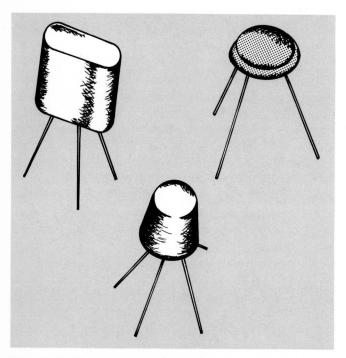

Fig. 18-5. The transistor is the basis for many electronic circuits in current communication devices.

Modern electronics was made possible by the invention of the transistor. This electronic device can switch current on and off (as in a computer). It can amplify a signal (for radio or TV reception). Electronic devices often contain thousands of transistors. They are arranged in integrated circuits or on printed circuit boards, Fig. 18-6. In fact, some printed boards are so small, they are made on silicon or germanium crystals. We refer to these as SILICON CHIPS. Integrated circuitry and silicon chips have replaced many regular transistors in modern electrical devices.

EARLY COMPUTER TECHNOLOGY

Electronic computing started in the 1940s. Early computers required hand assembly and were quite large. The Electronic Numerical Integrator and Computer (ENIAC) of 1946 contained 18,000 vacuum tubes! Obviously, the machine filled an entire room. It was fairly slow and needed an internal cooling system. Still, the ENAIC caused a great deal of excitement in the electronics industry.

Computer technology is said to have progressed through five GENERATIONS, Fig. 18-7. The earliest models typically depended on vacuum tubes for electronic switching. This marks first generation computing. It gave way to transistor technology; the second generation. These machines of the early 1960s relied on the smaller, more powerful transistors. The reduced size greatly increased the calculating speed of the machine. However, the computer was still slow and complex by today's standards.

Third generation computers contained several distinct improvements. Machines built in the late 1960s included small silicon wafers, known as MICROPROCESSORS. Now, the entire "brain" of the computer could be reduced to a small chip. These printed chips replaced the wires and tubes of earlier devices. Fully electronic circuitry provided new possibilities for computer users. The most useful was the ability to program the machine. Up to this time, only one program at a time could be used or stored in the computer's memory.

Personal computers in use now are perfect examples of fourth generation, Fig. 18-8. These very

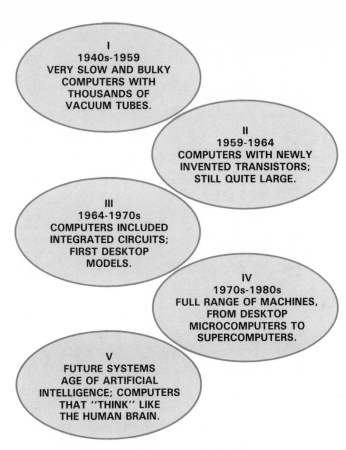

I
1940s-1959
VERY SLOW AND BULKY
COMPUTERS WITH
THOUSANDS OF
VACUUM TUBES.

II
1959-1964
COMPUTERS WITH NEWLY
INVENTED TRANSISTORS;
STILL QUITE LARGE.

III
1964-1970s
COMPUTERS INCLUDED
INTEGRATED CIRCUITS;
FIRST DESKTOP
MODELS.

IV
1970s-1980s
FULL RANGE OF MACHINES,
FROM DESKTOP
MICROCOMPUTERS TO
SUPERCOMPUTERS.

V
FUTURE SYSTEMS
AGE OF ARTIFICIAL
INTELLIGENCE; COMPUTERS
THAT "THINK" LIKE
THE HUMAN BRAIN.

Fig. 18-7. The five generations of computers.

powerful machines fit easily on a desk or tabletop. They operate much more efficiently and at a fraction of the cost of earlier models.

The term ARTIFICIAL INTELLIGENCE is closely associated with fifth generation computers. They are currently in the developmental stage. These computers are capable of making decisions and ethical judgements. However, there is some controversy about the need or value of these computers. This is just one question that advances in electronic communication force our society to answer.

EARLY HARD-WIRED SYSTEMS

By definition, a HARD-WIRED SYSTEM includes systems or equipment permanently connected by wire. The telephone is a common example. Most household telephones are linked to millions of other telephones by wires. This same concept is also true of telegraph and teletext devices.

Original development of the telephone was made famous by Alexander Graham Bell. Bell became fascinated by the study of electricity while a speech student in the 1860s. His experiments with tuning forks "sparked" several ideas in his head.

Fig. 18-8. This desktop computer takes up very little space and can deliver the power of a mainframe computer. (A.B. Dick Company)

Most notably, he wondered if two forks connected with a wire could transfer audible sounds between two remote points. Eventually, Bell started work on a device to send and receive the vibrations (or tones). By 1876, he publicly demonstrated the first telephone to the world, Fig. 18-9.

The telephone industry has improved dramatically over the years. Modern switching machines help route thousands of telephone conversations at one time, Fig. 18-10. In addition, the term "wire" does not fully apply to many calls now placed along communication lines. Instead, a typical telephone call today may travel through an orbiting satellite instead of across land wires.

The most noticeable changes, however, in modern telephones are in the home unit itself, Fig. 18-11. Features like multiple lines and portable units are fairly recent improvements. Some models have a memory for frequently-called numbers. Pushing one button automatically "dials" a friend or an emergency number.

ADVANCES IN LIGHT COMMUNICATION

Light has long been a medium of human communication. From the earliest camp fires, we have utilized visible light for signaling others. Lighthouses were originally built to warn ships of dangerous areas along shorelines. An entire system of railroad signals has directed trains since early days. And, perhaps you remember reading about Paul Revere signaling from the Old North Church?

Fig. 18-10. Modern telephone switching machines route conversations quickly and efficiently.
(Rockwell International Corp.)

Fig. 18-9. While early telephones were quite large and inefficient machines, their invention changed the history of communication.

Fig. 18-11. A typical home telephone unit can include a variety of options: second lines, call waiting, and automatic redial.

Modern transportation would be impossible without light communication. When riding in a car, count the number of traffic lights you pass. Sea and air travel require light signals, too, Fig. 18-12. Can you think of several examples of each signaling system?

Recently, light has been used in a slightly different form. Many telephone conversations now travel as a series of light signals, instead of as electrical pulses. This is called FIBER OPTICS, Fig. 18-13. Instead of using copper wires, signals are sent through materials such as glass or quartz fibers. Rather than an electrical current, small pulses of light are flashed along the wires. These light signals can be transmitted much faster along fiber cables than can electrical current along a copper wire. In addition, glass fiber lines are capable of carrying thousands of messages at once, instead of a single electrical pulse typical in other systems.

ADVANCES IN ACOUSTICAL COMMUNICATION

During our lives, we will receive a great deal of information through sound. Conversations with friends are dependent upon the exchange of sound waves, Fig. 18-14. Listening to popular music is also dependent upon sound waves. In both of these examples, the sound waves are types of acoustical communication.

Undoubtedly, the earliest form of acoustical communication started with the grunts of cave dwellers. Languages and musical instruments soon

Fig. 18-12. Visual communication is especially important around airports. Can you identify how light signals are used in this scene?

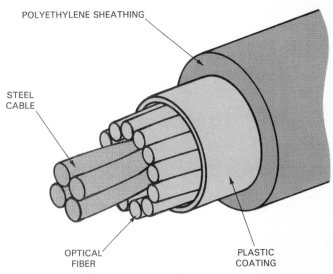

POLYETHYLENE SHEATHING

STEEL CABLE

OPTICAL FIBER

PLASTIC COATING

Fig. 18-13. Fiber optic cables are made of crystal clear glass or quartz fibers.

Fig. 18-14. Human conversation is the most common form of acoustical communication. (Ball State University)

added to the sounds created by humans. Technical systems of acoustical communication started with the clicking sound of the telegraph.

A great deal of acoustical information travels as INAUDIBLE SOUND. This means the wavelengths are outside the normal range of hearing, Fig. 18-15. Radar (RAdio Detection And Ranging) and sonar (SOund NAvigation and Ranging) are examples of inaudible sound. With these systems, sound waves are transmitted and received by electronic equipment.

Modern radar technology uses echoes to locate distant objects, Fig. 18-16. It was originally developed to warn of air attacks during World War II. Radio waves were "bounced" off approaching planes, long before they became visible.

Sonar systems are used in water. They are useful for identifying distances to the bottom of lakes or oceans. With sonar, pressure waves (sound waves) are used instead of radio waves (radar). This is because radio waves would be absorbed by the water.

MODERN ELECTRONIC COMMUNICATION

The next six chapters of this book explore electronic communication, and its various systems, in greater detail. Each of these chapters will explore a specific use of electricity for communication. You will discover how waves of electrical energy are used during the communication process. And you will also discover how they are changed to usable forms. First, however, let us go through a brief introduction to the content in these chapters.

HARD-WIRE SYSTEMS

A vast majority of modern electronic communication systems are still connected by wires. Transmission lines for telephones and telegraphs are typical examples. The internal circuitry of stereos and computers are also hard-wired. Your school P.A. system is connected by wires to the main office.

The first method of electronic communication sent pulses along wires. Even today, wires (and larger cables) are important mediums of transferring signals, Fig. 18-17. Most wiring used in elec-

INAUDIBLE SOUNDS

AUDIBLE SOUND

Susan!!

Fig. 18-15. Audible sounds are noticeable to the human ear. Waves of energy that are not noticeable are inaudible sounds.

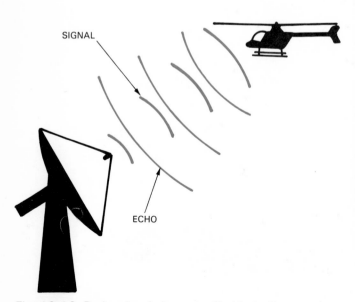

SIGNAL

ECHO

Fig. 18-16. Radar signals bounce off objects like an echo.

Fig. 18-17. Large electronic devices often include miles of wires. Inspectors must carefully check each wire before installing the equipment. (Zenith Electronic Corp.)

tronics is made of copper. Silver, tungsten, brass, and other conductors are also used in wires and cables. Large underground cables often include a combination of metals, called alloys.

TELECOMMUNICATIONS

Have you ever noticed all the antennas and satellite dishes in your community? Large radio towers dot the countryside and rooftops of our nation. These are examples of the latest technology in telecommunications, Fig. 18-18.

TELECOMMUNICATION generally refers to any exchange of messages over long distances. It is much more than a local telephone call to a neighbor. Modern telecommunication systems operate around the globe.

Today, the field of telecommunications typically includes the broadcast and information technologies. This includes radio, television, computers, and cable TV. On a smaller scale, walkie-talkies and CB radios are a form of personal telecommunication.

The most important characteristic of telecommunication is its ability to transmit messages to another location easily and quickly. Permanent wires or cables are not required between stations and receivers. Portable radios and TVs can pick up stations almost everywhere. Newscasters can go directly to the scene for remote location reports. With a single satellite, anyone is in touch with nearly half the globe, Fig. 18-19.

Fig. 18-18. Microwave antennas and satellite dishes are familiar sights in many communities.

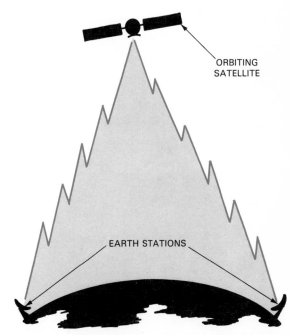

ORBITING SATELLITE

EARTH STATIONS

Fig. 18-19. Information is easily beamed around the globe by bouncing signals off orbiting satellites.

Introduction to Electronic Communication 177

LIGHT COMMUNICATION

Lasers and fiber optics highlight current systems of light communication. However, we should not ignore traditional methods of communicating with visible light, Fig. 18-20. These techniques are far from outdated. We still enjoy watching motion pictures and seeing brightly lit signs.

Light waves are a form of energy. Producing light that is visible to the naked eye is fairly simple. This type of light is called INCOHERENT. It includes a variety of frequencies (wavelengths). Different "colors" of light are caused by these varying wavelengths. Other types of light, for instance, infrared and laser light waves are all the same wavelength. The entire beam is created by waves of a single frequency. We call this form of energy COHERENT light, Fig. 18-21.

The majority of light communication systems used today are formed by incoherent light. Good examples include control devices for transportation, traffic lights, and emergency lights on cars. The fluorescent bulbs in lighted store displays are also incoherent light.

ACOUSTICAL COMMUNICATION

The most unique characteristic of acoustical communication is the use of sound as the medium of transmission. These signals may or may not be detectable to the human ear. But whether or not they can be heard, they are still forms of acoustical communication.

The major systems of acoustical communication will be described in Chapter 22. Most of these will deal with audible signals. However, sonar and radar will also be explored. These forms of communication are being used more and more in our highly technological world.

COMPUTERS AND DATA PROCESSING

Computer technology allows humans to exchange and supply large amounts of information in many forms. Large or small, computers have tremendous potential in handling vast amounts of data, Fig. 18-22. They can add daily sales to a

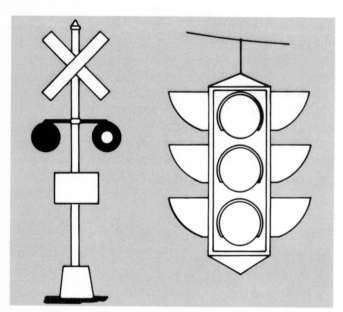

Fig. 18-20. Traffic signals represent an important type of light communication.

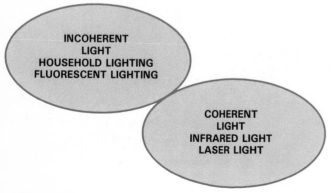

Fig. 18-21. Coherent and incoherent light waves.

Fig. 18-22. A typical business computer system for data management and processing. (Sperry Corporation)

yearly total or compute a difficult engineering equation.

Computer equipment can be divided into very simple systems, Fig. 18-23. INPUT DEVICES are a means for putting data into the computer. The keyboard on a microcomputer is a common input device. Universal code readers at grocery stores are another type of input device. They "read" coded strips from packages with beams of light. The numbers are then fed into the computer memory of the cash register for recording.

The internal memory of a computer is called a CENTRAL PROCESSING UNIT (CPU). This electronic device can calculate and manipulate information at great speeds. Early CPUs often filled entire rooms and required large amounts of energy. With the invention of the silicon chip, computers have become smaller. And microprocessing CPUs are small enough to fit into wristwatches and small toys.

OUTPUT DEVICES provide the needed information in a usable form. A computer monitor (screen) is the most common type of output device. Words, numbers, and pictures appear on the screen for quick reference, Fig. 18-24. Information may be output to other pieces of hardware,

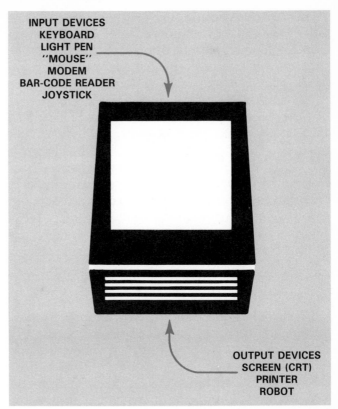

INPUT DEVICES
KEYBOARD
LIGHT PEN
"MOUSE"
MODEM
BAR-CODE READER
JOYSTICK

OUTPUT DEVICES
SCREEN (CRT)
PRINTER
ROBOT

Fig. 18-23. Examples of input and output devices on computer systems.

Fig. 18-24. Computer monitors and printers are two familiar types of output devices. Can you think of others? (Xerox Corp.)

Introduction to Electronic Communication   179

too. For example, data is easily "dumped" onto computer disks or magnetic tape. This information might then be sent directly to a printer. The printer will produce the data in printed form. Other messages may travel along phone lines to other computers.

Computers operate by following a series of instructions called PROGRAMS. Another familiar term for programs is SOFTWARE. Word processing software is useful when typing letters or reports. Mathematical software can perform calculations quickly. Graphics software helps in creating charts or pictures. Operation software is necessary just to "tell" the computer how to use other programmed instructions.

SUMMARY

Most of us are familiar with modern communicaton systems. Several of these systems rely on electricity in order to function. These include telecommunication, hard-wire, light, and acoustical systems. Computers and data processing systems are also important communication tools in our society. Each type of system was introduced in this chapter.

These electronic systems, as they exist today, developed over the course of many years. Probably one of the earliest electronic communication devices was the telegraph. And in the years since its development, electronic communication has progressed quickly. It is now even possible to send telephone messages with light waves.

The specific histories of each type of electronic communication system will be explored in the next several chapters.

KEY WORDS

All the following words have been used in this chapter. Do you know their meanings?

Acoustical energy, Artificial intelligence, Central processing unit, Circuit board, Coherent light, Computer, Electromagnetic energy, Electronic communication, Fiber optics, Generations, Hard-wired systems, Inaudible sound, Incoherent light, Input device, Microprocessor, Output devices, Programs, Radar, Radio waves, Silicon chips, Software, Sonar, Stereo, Telecommunications, Telephony, and Transistor.

TEST YOUR KNOWLEDGE—Chapter 18

(Please do not write in this text. Place your answers on a separate sheet.)
1. What is electronic communication?
2. List the five features of public radio broadcasts.
3. Which of the following is NOT a characteristic of transistors?
   a. They are powerful amplifiers.
   b. They are very large.
   c. They are inexpensive.
   d. All of the above.
4. _____ _____ are printed circuit boards made on silicon or germanium crystals.
5. Briefly explain each of the five generations in computer technology.
6. Define hard-wired system.
7. Inaudible sound is that which is detectable to the human ear. True or False?
8. What is telecommunication?
9. How do incoherent light and coherent light differ?
10. What does the term CPU stand for?

ACTIVITIES

1. Take apart an old radio. Identify the major components. It may be necessary to obtain a book from the library in order to help you with your research. Make note of any "unusual" parts.
2. Using a frequency generator, create different radio and sound waves. Discuss in class the various waves made and how each of them may be used.
3. Build your own telegraph system. Keep a list of the materials that were necessary in order to build the system. When your telegraph is complete, use it to send messages to your classmates.
4. Research how satellite dishes operate. Write a report on your research. Include photos of some of the different models, along with prices and the primary users of the different models.
5. Arrange to tour a local telephone switching station. After your tour, write a short paper explaining the route a typical phone call takes into and out of the switching station.

Basics of Electronic Communication Systems

The information given in this chapter will enable you to:

O *Discuss the basic concepts of electrical energy.*
O *Explain the types of signals used in electronic communication.*
O *Explain how to create code, transmit and receive messages using electronic devices.*
O *Discuss the technical equipment that aids many electronic communication systems.*

Electricity has been available to humans for a long time. Our world could not exist without electricity. When the first telegraph systems were made, electricity provided the power to transmit the messages over great distances. Now, computers, televisions, disc players, and radios all rely on electrical energy.

In order to understand electronic communication systems, one must first understand the terms used in this complex technology. The basics of electronic systems are covered in depth in this chapter. We will review basic concepts of electricity. Then, we will study the ways electricity is put to use in communications. This knowledge will help you better understand electronic communication systems, Fig. 19-1. They are discussed in the next few chapters.

Fig. 19-1. This microwave tower aids communication. Telephone calls, television and radio signals can be sent through it to a distant destination. (Rockwell International Corporation)

CONCEPTS OF ELECTRICAL ENERGY

The ATOM is the building block of all substances: land, water, and people, for example. Fig. 19-2 shows a model of an atom.

The center of an atom consists of a small particle called the NUCLEUS. Within the nucleus are PROTONS. They have a positive (+) charge. Also within the nucleus are some particles which have neither a positive nor a negative charge. These are known as NEUTRONS. Around the nucleus revolve smaller particles that have a negative (−) electrical charge. These particles are called ELECTRONS.

Generally, each atom has an equal number of protons and electrons. The positive charge from the proton cancels the negative charge from the electron. It is said that the atom is BALANCED, Fig. 19-3. The atom is held together by this force between the parts.

An atom will usually stay in a balanced state, unless some type of force or pressure is added to it. When energy is added, the atom becomes "excited." Any electrons loosely bound to the atom (those in outer rings) may be broken free. The atom, then, is no longer balanced. Electrons from other atoms are attracted to the unbalanced atom. The newly unbalanced atoms then attract still more electrons. This movement of electrons is known as ELECTRICITY.

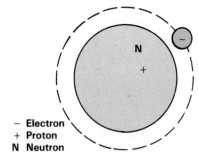

— Electron
+ Proton
N Neutron

Fig. 19-2. Study this model of an atom. Can you name the force behind each part?

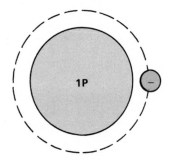

1P

Fig. 19-3. A hydrogen atom has an equal number of protons and electrons.

Some metal-like materials, such as copper and silver, allow easy movement of free electrons. These materials are called CONDUCTORS. But other materials do not allow easy movement. They are called INSULATORS. Several conductors and insulators are listed in Fig. 19-4.

As was stated, a force needs to be applied to produce electricity. It can be supplied by a battery or even a strong flame. It is called VOLTAGE. The actual flow of the electrons is known as ELECTRIC CURRENT. If there is no

CONDUCTORS	INSULATORS
Silver	Glass
Iron	Rubber
Brass	Dry Air
Copper	

Fig. 19-4. Common conductors and insulators.

flow, there is no current.

Electric current also produces a field of magnetic energy. This field is often called the ELECTROMAGNETIC FIELD. If this field is made to OSCILLATE (rapidly change direction back and forth), ELECTROMAGNETIC WAVES are created. It is with these waves that messages can be transmitted electronically.

Electromagnetic waves can take any number of forms. The signals sent out by your favorite radio station are electromagnetic waves. So is the X ray taken of your teeth by your dentist. Lightning that shoots across the sky during storms is another form of electromagnetic waves.

Typical electromagnetic waves are measured in several ways. As the wave passes a point (or object), the energy level changes or FLUCTUATES. This fluctuation is the result of high and low points formed in the wave, Fig. 19-5. The highest point is called a CREST. The lowest point is called the TROUGH. The distance between the crest and trough is referred to as the AMPLITUDE. The distance between the crest of one wave and the crest of the next wave is called the WAVELENGTH. It is a physical measurement. Finally, the number of waves passing a given distance in one second is the FREQUENCY of the wave. It is a timed measurement.

The major difference between waves is their frequency and length. Without varying wavelengths and frequencies, all electromagnetic waves would be the same. There would be no way to communicate.

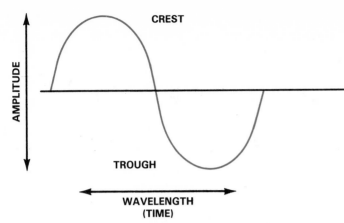

Fig. 19-5. The parts of an electromagnetic wave.

RESISTANCE

Resistance is the opposition to electrical current. It can be caused by a material, substance, component, or device. For example, resistance is found in every part of an electric current. Even the wires in simple circuits provide some resistance. This type of resistance cannot be avoided. It is caused by the actual make-up of the electrons in wires and electrical components.

There are times when it is necessary to vary the strength of the current. In this case, there are types of resistance that can be purposely placed in a circuit, to get a specific result. These RESISTORS divide voltage and/or limit current. For example,

a radio volume control is a form of resistance. It limits the amount of voltage that reaches the speakers.

ELECTROMAGNETIC SPECTRUM

All electromagnetic energy can be found in the ELECTROMAGNETIC SPECTRUM, Fig. 19-6. The human eye can see only a small portion of this spectrum. This is known as VISIBLE LIGHT. We cannot see beyond violet or below red in the spectrum. Other visible colors in the spectrum include orange, yellow, green, blue, and indigo.

You will notice there are many more waves in the electromagnetic spectrum than just visible light. GAMMA RAYS are the shortest waves in the spectrum. They are caused by radioactive decay. Next to gamma rays are X RAYS, which are slightly longer than their neighbor to the left. The shortest light that still escapes detection by the human eye (or for most) is ultraviolet light.

Those waves that are longer than visible light are known as the AUDIO SPECTRUM. Because they are longer than visible light, they cannot be seen by the human eye. The waves in this area can be heard. We often call these AUDIBLE SOUND WAVES.

The shortest of these longer waves is infrared light. This light is used for many functions. For example, it "does" the focusing on self-focusing

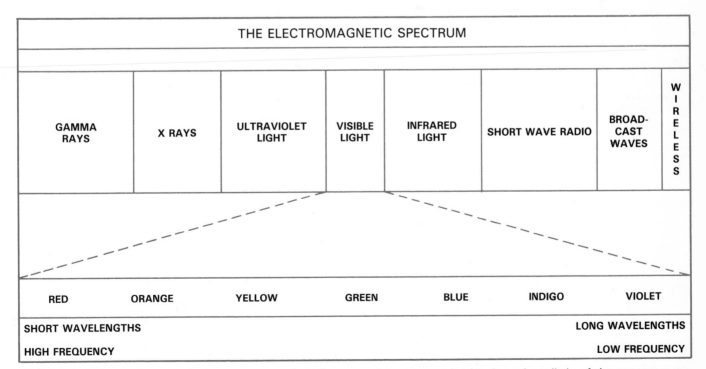

Fig. 19-6. The electromagnetic spectrum contains all waves. Are you surprised to learn how little of the spectrum we can actually see?

cameras, Fig. 19-7. Infrared light is also used with burglar alarms.

Next to infrared are the radio frequencies. Refer back to Fig. 19-6. You will notice that short wave radio waves are the shortest of the radio waves. The longest are "wireless" waves. In between short wave and wireless are the sound waves that accompany television broadcasts.

All of these waves are electromagnetic energy. They are voltages or currents that change with time. To communicate using electromagnetic waves, we must be able to send, receive, amplify, and modify them to serve our purposes. We must be able to use them as signals.

SIGNALS

A signal is a sound or image sent in a number of ways. It has a noticeable voltage or current that can be used to transmit information. Two signals used quite often in communications are digital signals and analog signals. They are used on recordings, telephone conversations, and computers, to name a few.

Analog signals

Analog signals vary constantly. They are smooth and continuous. Voices are analog signals; they change constantly, over a wide range. At times a voice can be soft. Other times a voice can be very loud. Analog signals can vary a great deal in amplitude and frequency.

Analog signals can be troublesome when being transmitted. Because a message varies in strength, weak signals, high points, and low points can be distorted by noise or other interference.

Despite this drawback, however, analog signals are useful. They allow us to hear every voice inflection. They allow us to see every shade of a color. Analog devices that transmit these varying signals add to the complete reception of a message.

Digital signals

Digital signals can represent only one of two states: on or off, 1 or 0, true or false, Fig. 19-8. In this form, digital signals are most often used in computers (calculators, personal computers, mainframes, etc.). This is known as BINARY CODE. A computer understands and communicates in this way. The number of on/off signals in a specific place represents a number, letter, or other character to the computer, Fig. 19-9.

Digital signals are also used in telephone transmissions. However, in this instance, they are converted from an analog signal to a digital signal. This change takes place in the following manner.

The sound wave from a voice is first converted into an analog signal. The analog signal is tested and sampled at certain intervals. Each of the points that was tested and sampled is given a value. This value is a combination of on/off pulses, or a digital signal. Refer to Fig. 19-10.

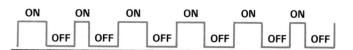

Fig. 19-8. Digital signals have only two states.

BINARY CODES	DECIMAL EQUIVALENT	ALPHABETIC EQUIVALENT
1000	8	S
1001	9	T
0100	4	O
0101	5	P

Fig. 19-9. This combination of 0, 1 digits represents numbers. The 0-1 figures are called binary codes.

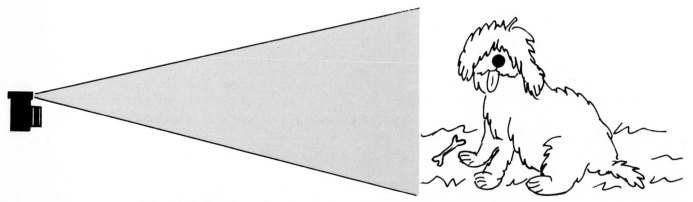

Fig. 19-7. Self-focusing cameras use infrared light to detect distances.

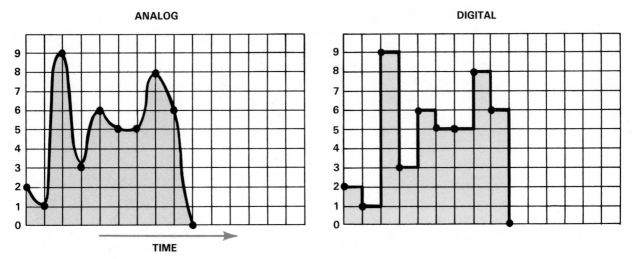

Fig. 19-10. The analog signal is sampled thousands of times each second.

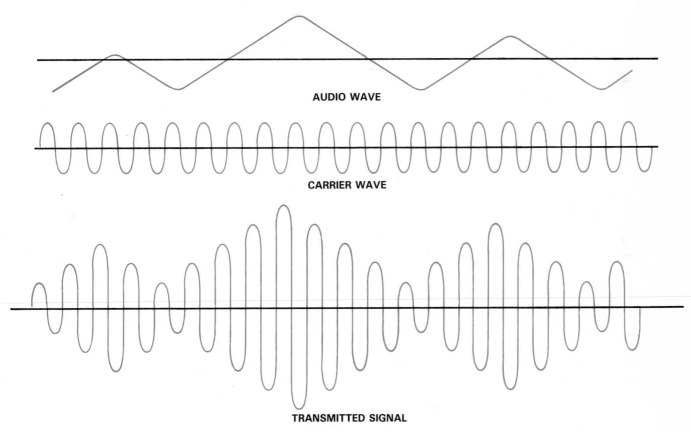

Fig. 19-11. An AM radio broadcast signal. The audio wave and the carrier wave are joined to make the signal that is transmitted.

BASIC ELECTRONIC COMMUNICATION SYSTEMS

All of the waves and signals discussed to this point would be useless if we lacked the technology to transmit them. The systems that are used to transmit and receive messages are the basis of modern electronic communication.

Three pieces of equipment are needed to send and receive messages. They are a transmitter, a transmission link, and a receiver.

TRANSMITTERS

A transmitter is used to start a message on its journey. Within the transmitter, the message is changed into electrical signals. The signals may be changed into analog waves or digital waves. At this point it can be sent directly to an amplifier. Or it may need to be modulated.

There are instances in which electrical signals need help in being transmitted. These signals are MODULATED.

MODULATION is the process of controlling a carrier wave. The signal is fed into a modulator. The modulator causes the radio carrier wave to change. It changes either in amplitude (AM) or frequency (FM).

Amplitude modulation

Amplitude modulation is used most often in basic AM radio broadcasts. The audio signal in an AM broadcast is often a voice or music. Audio signals in this case are changed into electrical signals by a microphone, tape deck, or turntable.

Such a signal is often quite low in voltage. Therefore, it must be amplified before it goes to the modulator. In addition, a carrier wave must be created. Both the information wave and the carrier wave are sent through the modulator. The resulting wave is shown in Fig. 19-11. The AMPLITUDE of the wave is changing.

At the destination of the message, a DEMOD-ULATOR is used. It sends out only the original audio signal. The amplifier in the demodulator boosts the signal to the power levels needed by the speakers. The speakers then deliver the sound sent originally.

Frequency modulation

Frequency modulation, or FM, is similar to AM. The basic difference is that the modulator works differently. In FM the modulator must vary the frequency of the wave, Fig. 19-12.

FM has a few advantages over AM. FM uses power more efficiently. Noise has little affect on FM. This improves sound quality of the transmission. FM radio signals are generally clearer because all the information is in the frequency of the signal. Noise usually gets caught up in the amplitude of a signal.

There is a second advantage to FM. When two signals reach a transmitter (antenna) at the same time, only the stronger signal will be heard. So

when you are listening to the radio in a car, you will hear the signal sent from the strongest and nearest transmitter. There will be no interference from another, more distant, transmitter.

TRANSMISSION LINKS

The next step in communicating a message is the transmission link. Its job is to relay the signals on to their destination.

The majority of communication signals are sent by cables or through the atmosphere. These methods allow for the best transmission of a signal. Generally, radio and most television signals are sent through the atmosphere. However, more and more television signals are being linked by cables and satellites. Also, optical fibers are becoming popular for linking telephone signals.

Antennas

Antennas aid us in receiving signals sent through the atmosphere. Basically, they function in the same way. The antenna picks up the oscillating current sent from the transmitter. This current is sent outward from the transmitting antenna as electromagnetic waves. Some distance from the transmitter, waves are picked up by an antenna attached to a receiver. Inside the receiver, special electronic devices receive and boost the signal back to a usable level.

Antennas are made in a variety of shapes and sizes. This is because each type of antenna is made to receive a certain type of signal. For example, a microwave antenna is used to receive signals from distant satellites or other microwave antennas. The dish portion is usually quite large. This size helps pick up distant signals, Fig. 19-13.

Other types of antennas include whip antennas (radio) and yagi antennas (television), Fig. 19-14.

Cables

Another method for linking signals is through cables. Cables allow the transmission to be totally controlled. This is because the signals are sent along wires enclosed in a metal covering. The covering keeps the signals going in the correct direction.

For signals in very-high frequency (VHF) and

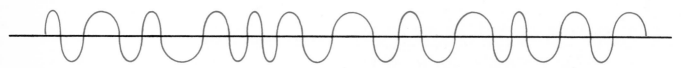

Fig. 19-12. In an FM signal, high frequency is used for high-amplitude messages, low frequency for low-amplitude messages.

Fig. 19-13. Microwave relay towers must be in line of sight of each other. This is because microwaves travel in straight lines.

lightning, thunder, and even the sun can cause noise.

Devices such as optic fibers, cable, and digital communications reduce the noise in transmission. These devices are made in such a way that noise is not introduced.

This one factor must be accounted for when sending a message. While a message may be sent in several ways, the method with the least noise will usually be chosen. A clear message is a very important part of the communication process.

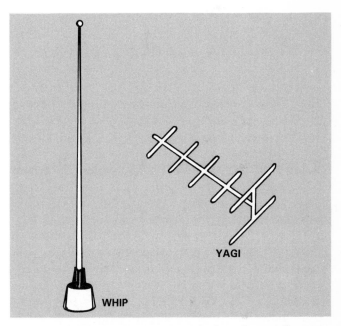

Fig. 19-14. The yagi antenna works in only one direction. Whip antennas operate in many directions.

lower ranges, coaxial cables are the best transmitters, Fig. 19-15. The cable itself is a wire that is surrounded by insulating material. This material holds the wire in place. The wire and insulation are then enclosed in the metal covering already discussed.

When sending signals in the infrared and visible light range, another type of cable is used. This cable is made up of a glass fiber, or an OPTIC WAVE GUIDE. The transmission is known as FIBER OPTICS. It is very effective for linking very-high frequency messages a long distance.

Fig. 19-16 illustrates the construction of an optic fiber cable. The diameter of the fiber is very small. This reduces the chance for interference. The transmitting fiber is then surrounded by an energy absorbing glass. This glass will absorb any rays that escape the transmitting fiber. The transmitter and receiver are designed to work at very high speeds. This means messages can be exchanged quickly.

Noise

Noise is any signal not present in the source message. It is usually picked up in the linking stage. This is because the signal often goes through the atmosphere at this point. For example, an AM broadcast is likely to attract noise. The noise is picked up when the message leaves the transmitter on its way to the receiver. At that point, the signal is moving through the atmosphere. Rain,

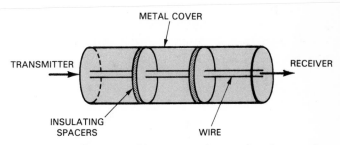

Fig. 19-15. Coaxial cables are made up of an inner wire, surrounded by three layers of material.

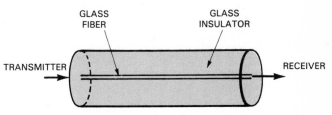

Fig. 19-16. Optic fibers are a high-speed link between a source and a destination.

RECEIVERS

The receiver is the final electrical stop for a message. Here it is detected and converted back to the source message. Fig. 19-17 shows the complete journey of a message.

There are many electronic devices that receive signals. The receiver in a stereo system does just what the name implies. It changes an electronic wave back to a voice or music. Similar receiving units can be found in a television set, a radar system, and a computer terminal.

You will learn much more about receivers in specific areas in the next several chapters.

SYSTEM FUNCTION AND CONVERSION

Additional functions and conversions occur during the trip taken by a message. These processes happen in particular devices. With these components, electronic communication is possible.

One device that aids in this process is the amplifier. An AMPLIFIER increases the power of the signal, or amplitude. It is used on weak signals. By amplifying the signal, it becomes powerful enough to be used. For example, the amplifier in a stereo system boosts signals from a record, tape, or radio. It is then at a level that can activate the loudspeakers.

Another device that aids the transmission of a message is a CONVERTER. A converter in an AM radio changes incoming signal frequency to a frequency that can be detectable. Once it is detected, it can be amplified, then converted back to audio.

A third device that aids the electronic communication process is the OSCILLATOR. This device produces repeating signals. These signals are of a certain frequency and amplitude. It is a special type of amplifier. It makes a constant signal

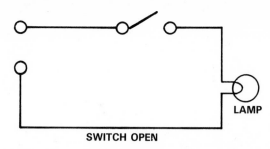

Fig. 19-18. This switch directs the electricity to a lamp.

in analog and digital systems. This signal is what is added to the original signal in radio broadcasts, to make the modulation.

In AM broadcasts, an oscillator produces a signal that varies in amplitude. In FM broadcasts, it is the frequency that is varied by the oscillator.

SWITCHES direct the flow of electricity, Fig. 19-18. They are very basic to electronic communication. Switches send electricity to its destination, which allows the message to reach its destination also.

SUMMARY

Many modern communication systems rely on electricity in order to function. If we want to learn about these systems, we must first understand the theory of electricity.

Electricity is the movement of electrons. The electron movement is caused by a force: heat, friction, or even light. This flow is known as electric current. Resistance is the opposition to electric current.

Electricity is the basis for both the visual and audio spectrum. These spectrums can be used in the form of signals. Signals are used to communicate.

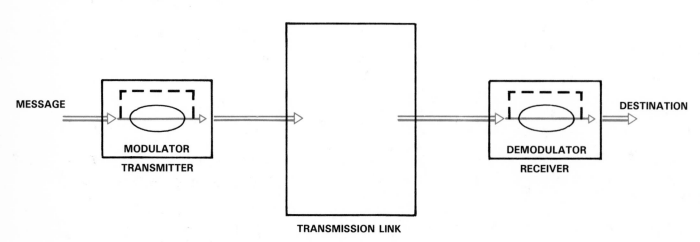

Fig. 19-17. A message travels through these channels to reach its destination.

Basic electronic communication systems use signals to communicate. To send a complete message, a transmitter is needed to send the message on its way. A transmission link relays the message to its destination. And, finally, a receiver accepts and decodes the message.

KEY WORDS

All the following words have been used in this chapter. Do you know their meanings?

Amplifier, Amplitude, Amplitude modulation, Analog signals, Atom, Audible sound waves, Audio spectrum, Binary code, Conductors, Converter, Crest, Digital signals, Electric current, Electricity, Electromagnetic spectrum, Electromagnetic waves, Electron, Fiber optics, Frequency, Frequency modulation, Insulators, Neutron, Noise, Nucleus, Optic wave guide, Oscillator, Proton, Receivers, Resistance, Switches, Transmission links, Transmitter, Trough, Visible light, Voltage, and Wavelength.

TEST YOUR KNOWLEDGE—Chapter 19

(Please do not write in the text. Place your answers on a separate sheet.)

MATCHING QUESTIONS: Match the definition in the left-hand column with the correct term in the right-hand column.

1. __ Movement of an electron from one atom to another.
2. __ Positively-charged particle.
3. __ Materials that allow the easy movement of free electrons.

    a. Nucleus.
    b. Conductors.
    c. Electric current.
    d. Atom.
    e. Electricity.
    f. Proton.

4. __ The building block of all substances.
5. __ The center of the atom.
6. __ The flow of electrons.

7. X rays, lightning, and stop lights are examples of _____ waves.
8. Explain the concept of resistance.
9. Name two types of signals used often in electronic communication.
10. Define modulation.
11. Amplitude and frequency modulation both begin in the:
    a. Receiver.
    b. Transmission link.
    c. Transmitter.
    d. None of the above.

ACTIVITIES

1. Make a model of an atom, using styrofoam balls and pipe cleaners. Label and explain each part.
2. Consult the Periodic Table to learn the structure of well-known elements such as gold, silver, and sodium. Choose one element and explain its structure. File a report with your teacher concerning your findings.
3. Invite a radio engineer to visit your class to discuss this job position. Ask your guest to explain what duties a radio engineer has. Learn what type of schooling or training is needed.
4. Make a chart with the headings Transmitter, Transmission link, and Receiver. As a class, list examples of each. Discuss how these examples work.

20 Telecommunications

The information given in this chapter will enable you to:
○ *List and explain each of the FCC categories.*
○ *Discuss the frequency bands used in telecommunications.*
○ *Explain the types of telecommunication systems.*
○ *Discuss the workings of a data communications system.*

In Chapter 5, we learned that TELECOMMUNICATIONS means transmitting information between distant points. Telephone networks are a good example of telecommunications. Many other systems are part of this industry. Satellites are used to send radio messages around the world, Fig. 20-1. Cable television allows students in Seattle to watch a session of Congress in Washington, D.C. Microwave networks provide a link between banks, stores, and other businesses.

Telecommunications is still an area of rapid change. A "state-of-the-art" product is often outdated in only a few months. This is due to the constant increase in understanding of this technology. As scientists, computer engineers, electrical engineers, and others involved in this industry learn more, telecommunications improves and changes. We are constantly finding new and better ways for communicating between distant points.

USES AND PURPOSES OF TELECOMMUNICATION

Chances are that sometime during each day you use a telecommunication system. When you call a friend on the telephone, you are using telecommunications. Your call is sent through a series of switches until it is connected with the telephone at your friend's house. When you listen to the radio or watch television, you are also using telecommunication systems.

If you did not have these methods of communicating, how would you react? You might feel out-of-touch and alone. We have come to depend on the convenience of communicating quickly and easily, over any distance.

In this way, telecommunications serve two purposes. First, it makes possible the quick exchange of messages over long and short distances. And second, it can be used in a number of ways. Almost any information can be coded, transmitted, and received in electronic form. If you need to send a photograph to a business associate in Germany, it can be sent through a telecommunication system as a photograph.

MODERN COMMUNICATION SYSTEMS

There are many types of telecommunication systems. And there are also many users of these systems. Each of these users requires a frequency on which to broadcast. Because of the large number of users involved, some regulation (control) is required. In the United States, regulation is the responsibility of the Federal Communications Commission (FCC).

The FCC has divided telecommunication systems into categories. Each of these categories can be used as shown in Fig. 20-2. In most cases, the user must get a license to operate.

FCC CATEGORIES

The major FCC categories are broadcasting, citizen communication, industrial and government communication, and commercial transportation communication.

Broadcasting

Broadcasting includes AM and FM radio and television transmissions. Television and radio

Fig. 20-1. Satellites make quick worldwide communication possible. (RCA)

FCC CATEGORIES	
Broadcasting	AM FM Television Links Common Carriers
Citizen Communication	Amateur Citizen Band
Government and Industrial Communication	Armed Services Government Departments Public Safety Industrial Communication Meteorological Services Telemetry Industrial, Scientific, and Medical Equipment
Commercial Transportation Communication	Paging Services and Car telephones Mobile Land Vehicles Aeronautical Control Aviation Maritime Navigational Beacons

Fig. 20-2. Federal Communications Commission categories.

broadcast services are used widely every day. We are all familiar with them. The frequency ranges of broadcast stations are shown in Fig. 20-3. There are also broadcast frequencies assigned for actually operating broadcast stations. These frequencies are normally only used by engineers in the broadcasting industry.

Broadcasting is perhaps the most common of the FCC categories. It uses many smaller, basic telecommunication systems, like the radio and television, to make up a completely new industry. The broadcasting industry will be discussed more fully in Chapter 23.

Citizen communication

Citizen communication consists of two areas: citizen band and amateur band transmissions.

These bands are used by citizens for personal communication. CITIZEN BAND frequencies are used for short distance communication and remote control products. Examples include CB radios, garage door openers, and television remote controls.

The AMATEUR BANDS are used for long-distance personal communication. These frequencies allow more options than CB bands. For example, amateur bands allow the user to send transmissions around the world. This can be helpful during emergency situations. During these times, radios are set up as an emergency communications network. They are staffed by civilian citizens.

Industrial and government communication

Industrial and government communication frequencies are for use by these groups. They are used in local and global communication of the government, armed services, and weather services.

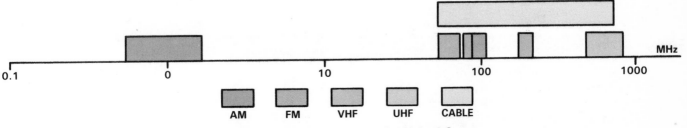

Fig. 20-3. Broadcast bands in the United States.

Local businesses and factories also use these frequencies during a regular business day. Using radios on these frequencies provides a link between the company and its business cars and trucks.

Commercial transportation communication

These frequencies are used by taxi, ambulance, and ship radios and also by paging services. Frequencies are also set aside in this category for coded electromagnetic waves. These waves are used to guide the direction of aircraft and ships. One system is known as the Instrument Landing System, or ILS. All major airports are equipped with this system.

FREQUENCY BANDS

Each of the FCC categories operates within a certain group of frequencies. The terminology (names) used for each of these groups of frequencies is shown in Fig. 20-4. We will review several of the frequency bands briefly.

The extremely low frequency (ELF) and voice frequency (VF) bands cover frequencies from musical instruments to voices. Very low frequency

TERMINOLOGY OF FREQUENCY BANDS		
Band Number	Frequency Range	Frequency Subdivision
2	30-300H	ELF Extremely low frequency
3	300-3,000H	VF Voice frequency
4	3-30kH	VLF Very low frequency
5	30-300kH	LF Low frequency
6	300-3,000kH	MF Medium frequency
7	3,000-30,000kH	HF High frequency
8	30-300MH	VHF Very high frequency
9	300-3,000MH	UHF Ultra high frequency
10	3,000-30,000MH	SHF Super high frequency
11	30,000-300,000MH	EHF Extremely high frequency
12	300,000-3,000,000MH	

Fig. 20-4. Study this chart of frequency terminology. Where do AM and FM broadcasts fit in the chart?

and low frequency have very long wavelengths. Huge antennas must be used for transmission. This is not very practical, so they are used only in special communication systems.

Medium frequency (MF) and high frequency (HF) are used for commercial AM broadcasting and also for short wave and amateur broadcasts.

When a transmission from these frequencies is sent out, it hits a layer of the earth's atmosphere, called the IONOSPHERE. The ionosphere reflects the signal back down to the earth. This signal may reflect back and forth any number of times. It is an action similar to that which occurs when a ball is bounced in an area that has a low ceiling. Sooner or later the signal will contact a receiver, Fig. 20-5. These reflections make transmission over a great distance possible.

Very high frequency (VHF) and ultra high frequency (UHF) are two terms with which you may be familiar. These bands are used in short distance, line-of-sight communications. In other words, receivers and transmitters must be directly in sight of one another. This limits the broadcast range of most TV stations.

Microwaves include all frequencies above the UHF range. The term microwave comes from the short ("micro") nature of these wavelengths. Radar communication is one common use of microwaves.

Because these microwaves are so small, they have several advantages. One advantage is that the signals can be beamed into a very small area. This means the energy is used efficiently. And because of this intense beam of energy, antennas can be very small in microwave use.

However, there are also disadvantages with microwave signals. Perhaps the biggest drawback is the effect of poor weather on microwaves. Because the wavelengths are so small, rain, snow, hail, or even a high wind can absorb the energy in the signal.

TYPES OF TELECOMMUNICATION SYSTEMS

The categories and frequencies we have discussed so far are only part of the area of telecommunications. In order for telecommunications to be useful, it is necessary to have devices to receive and transmit signals. These devices make a complete communication system.

Very likely you are familiar with several systems we will discuss, Fig. 20-6. Perhaps a few are new to you. In the following sections we will discuss how and where each of these systems is used in modern telecommunications.

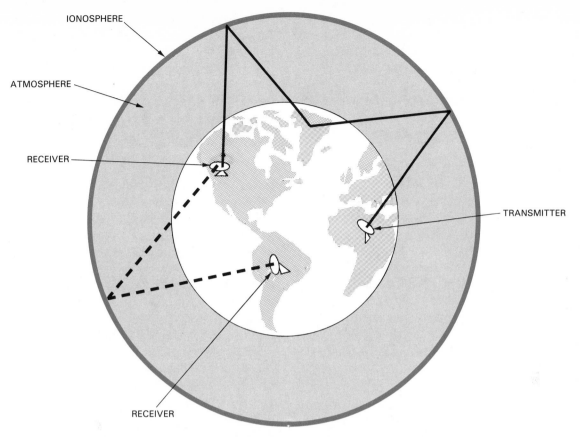

Fig. 20-5. MF and HF signals are sent out, where they hit the ionosphere. They are then reflected back down to a receiver.

Fig. 20-6. Many types of telecommunication devices are available. (Ball State University, Zenith Electronics Corp.)

RADIO

Radio is one of the most widely used telecommunication systems. We use it every day. Common uses include AM and FM broadcasts. But there are other ways we use the radio that we may not think of as radio use. One example is a car telephone which works using radio waves, Fig. 20-7. Another example is a remote controlled model airplane. It also uses a certain radio frequency to operate.

Radio messages can be transmmitted using one-way or two-way radios. ONE-WAY radios only send messages or receive messages between a source and a destination. There is no feedback involved. AM and FM broadcasts are one-way. Other one-way radio messages include pager systems or "beepers," weather band radio broadcasts, and garage door openers. The frequencies of these devices vary. They are found all along the frequency band. Most, however, can be found along the MF, HF, and UHF bands. Refer back to Fig. 20-3.

Radio messages can also be sent and received using TWO-WAY radios. With two-way radios, it is possible to have feedback. We can respond directly to messages on two-way devices. Taxi radios and CB radios are examples of two-way radios. When a taxi driver receives a message from a dispatcher (operator), he or she can respond to the dispatcher directly.

The telephone is also an example of a two-way radio. It has one major difference, however, from most other two-way radios. With a telephone, two signals can be sent at the same time. For example, suppose you and a friend are talking to each other on the telephone. If you both speak at the same time, both voices will be heard. However, with a taxi radio, only one person can talk at a time. This is because when speaking on a taxi radio, walkie-talkie, or a police radio, a "talk" button must be pressed. When it is pressed, it is impossible to receive a message, Fig. 20-8.

Radio signals are the most commonly used signals. They are used in mobile telephones and television broadcasts to name a few. While these are individual telecommunication systems on their own, they depend on radio waves to send their messages.

TELEVISION

All television broadcasts, including cable, can be found in the frequency bands ranging from

Fig. 20-7. Car telephones use radio waves to transmit signals.

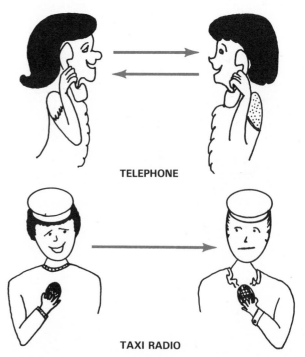

TELEPHONE

TAXI RADIO

Fig. 20-8. Examples of one-way and two-way radio systems.

VHF through UHF. VHF and UHF signals are sent through the atmosphere, while cable signals are transmitted through coaxial cables, Fig. 20-9.

The FCC has assigned specific channels for each television broadcast. Broadcasts in the VHF band are given channels numbered 2 through 13. The UHF band consists of channels from 14 and higher. Neither VHF nor UHF signals reflect a great deal off the ionosphere. They cannot travel as far as MF and HF waves. VHF and UHF bands transmit line-of-sight, short distance broadcasts.

Transmitting broadcasts through cables solves the problem of short distance broadcasts. The transmissions are made in the same way as telephone transmissions, Fig. 20-10. The studio sends out television broadcasts to each SUBSCRIBER (household that rents the service). These broadcasts are sent down a main line which has feeder lines tied into it. Each feeder line is connected to the lines that go to each subscriber.

As we learned in Chapter 19, interference does not affect a cable transmission a great deal. Also, cables offer a more focused signal. With the protective covering surrounding the cable, signals are directed without losing much power. This means a cable-transmitted broadcast will be clearer and will travel farther than a broadcast sent through the atmosphere.

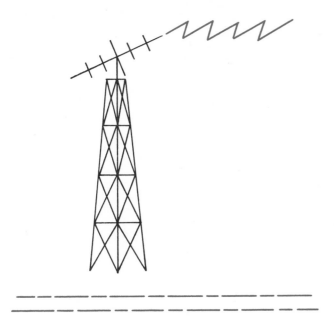

Fig. 20-9. Television signals can be sent through the atmosphere or cables.

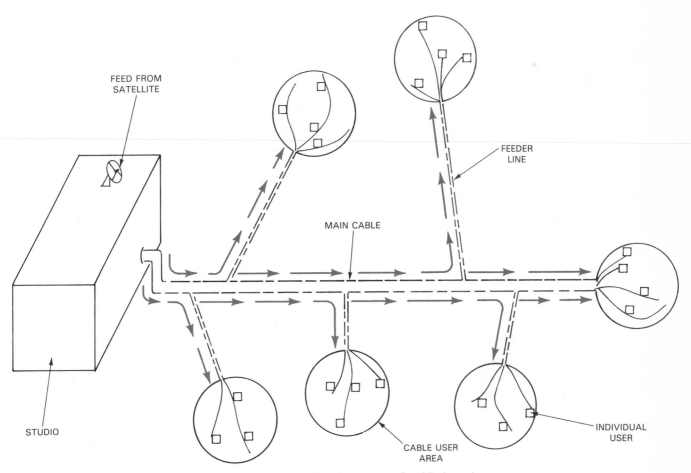

Fig. 20-10. The transmission route of cable broadcasts.

Once these transmissions are sent, however, how are the images reproduced (made) on the television screen? The picture is made by light that constantly moves and changes in levels of brilliance (lightness, darkness, color). These changes occur so rapidly the human eye cannot detect them.

A television contains a CATHODE RAY TUBE (CRT), Fig. 20-11. The CRT contains an electron gun that shoots electrons out onto the television screen. The screen is divided into 525 (or more) lines. Each of these lines is scanned every time the flow of electrons changes. The screen is coated with phosphors that light up when the electrons hit them. Depending on the image being sent from the television studio, the electron beam will change according to the strength of the signal. Now imagine this process taking place 30 times a second over the lines on the screen! It is now clear why the human eye cannot detect the movement.

Whether or not the picture is black and white or color depends on the picture tube. The process just explained is typical of a black and white picture. In the screen of a color television, there are three electron guns; one each for blue, yellow, and red. The screen is coated with three kinds of phosphors. When the electron beam contacts the phosphors, some glow red, some glow blue, and some glow yellow, Fig. 20-12.

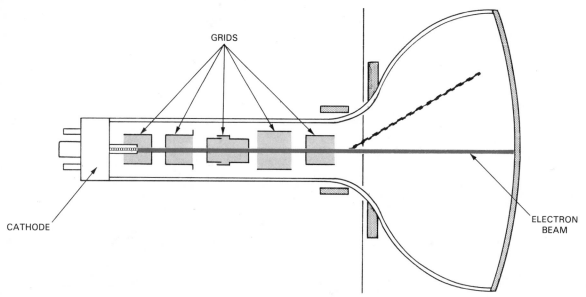

Fig. 20-11. Basic black and white television picture tube and screen. The cathode shoots an electron beam through a series of grids. These grids speed up and focus the electrons.

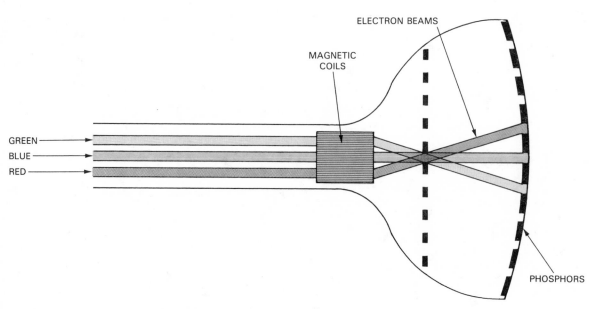

Fig. 20-12. Color TV picture tube and screen.

Television has grown a great deal since its beginning. With the constant improvements and innovations (new ideas), television plays a very important part in the telecommunications industry.

TELEPHONE

The telephone is an excellent example of telecommunications. It allows us to communicate over long distances quickly. Quick and efficient service is very important in telecommunications.

Quick service is supplied by telephone SWITCHING SYSTEMS, Fig. 20-13. Telephone switching systems route calls from the source to the destination.

In the early years of telephone use, switching was done manually (by hand). Operators connected the wire from the telephone line making the call to the telephone line of the person being called. But as telephone use grew, automatic systems came into use. Until recently, the major portion of switching was ELECTROMECHANICAL. That is, an electronic impulse (number dialed) caused the switching to occur.

While electromechanical switching is still used, most switching is now done through electronic switching systems (ESS). In these systems, programs are used to do routine operations, like switching. These programs are stored in computer memory. When a friend dials your number at home, the electronic pulses from dialing are collected. After all the digits have been dialed, the computer reviews them. It then maps out the quickest switching route to complete the call.

Electronic switching systems are far more efficient than electromechanical switching systems. ESS are quicker, less costly, smaller, more reliable, and use less power. The computers not only complete calls, they also control the overall function of the system.

Switching systems belong to a NETWORK. This network is worldwide and provides uniform telephone service to everyone. Phone service in Italy is able to work with the system used in the United States. If you wish to call a relative in Italy, there will be no problem with the two phone systems working together.

Fig. 20-13. This switching system is used for long distance service. (Rockwell International Corp.)

The telephone network is made up of transmission links, terminals (telephones), switching operations, and more. All these pieces work together, and are necessary for completing a call. Transmission links, for example, transmit the call, Fig. 20-14. Network switching systems complete the circuit. Telephones allow you to hear and be heard, Fig. 20-15.

TELETEXT AND VIDEOTEXT

Teletext and videotext systems are two methods for obtaining data stored in computers all over the country. Teletext systems are able only to receive information. Videotext, however, is active. The user can request information and act on information received. Both these systems use a television set, equipped with special attachments and modified to receive special signals.

Teletext transmits words and graphics (charts, drawings) through regular television signals. The user requests a certain page of information. It is then displayed on the television screen, Fig. 20-16.

This page of information is picked from all the information stored in the computer. It is similar to reviewing one page of an entire collection of library books.

Once the page is accessed, you cannot interact (talk) with the computer any further. Therefore, using our previous example, you can only read the page, perhaps even take a few notes on the data. You then may request another page of information. Teletext systems provide basic information.

Videotext systems do, however, allow the user to talk with the computer. Videotext signals are transmitted through telephone lines. The user interacts with the computer through a control device attached to the television set. Like the teletext, the data requested is displayed on the television screen.

A very good example of a videotext system is that used by visitors information centers. These systems can be used to find restaurants and points of interest in a particular area. Some stores also use videotext systems as a store directory. By answering a series of questions at a computer, you

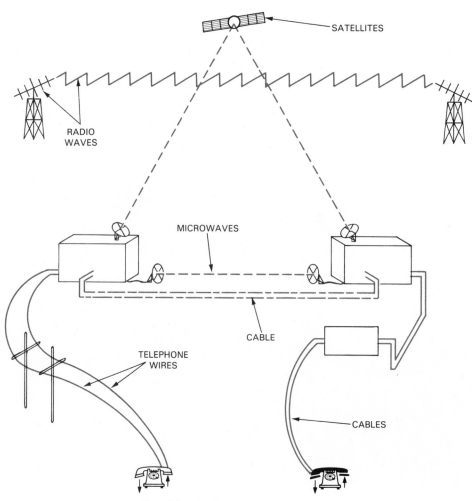

Fig. 20-14. Transmission options.

Fig. 20-15. Telephones allow messages to reach their destination. (AT & T)

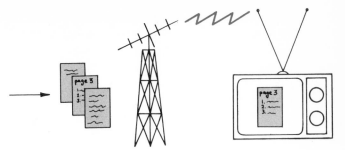

Fig. 20-16. How a teletext works.

are directed to the area of the store you seek.

Future uses of videotext include shopping from home and banking from home, both through a computer. You may even be able to read the evening newspaper using videotext! Videotext and teletext contribute a great deal to the Information Age.

SATELLITES

Satellites are truly a space-age device, Fig. 20-17. Development is due to advances made in such

Fig. 20-17. An orbiting satellite. (Ford Aerospace & Communications Corporation)

areas as space research, electronics, and robotics. Satellites are used as tools, as broadcasting devices, and as military communication systems. In order to understand satellites fully, we must understand basically how they work.

A communication satellite is designed to receive and send signals, Fig. 20-18. These signals must travel a great distance from ground stations on the earth's surface to the satellite (UPLINK) and back to the earth (DOWNLINK) again. Microwave beams are used for these transmissions. Microwaves are able to cut through the earth's atmosphere easily. The beams must be carefully aimed toward the proper ground station or satellite. The signals are sent and received using special antennas. The ground antenna and the satellite must be able to stay in constant contact with each other in order to transmit. If the satellite moves faster than the earth revolves, the contact would be broken.

Satellites must circle the earth at the same speed the earth revolves on its axis, every 24 hours. They will move at this rate when placed into orbit at 22,300 miles above earth. This is known as GEOSYNCHRONOUS ORBIT. At this point, a

satellite downlink covers 40 percent of the earth's surface. That means that signals transmitted from one earth station can be received by another earth station 8000 miles away! A message sent from Boston, Massachusetts, therefore, could reach a

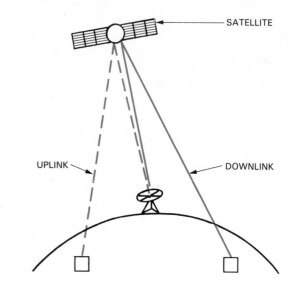

Fig. 20-18. The travel route of a satellite transmission.

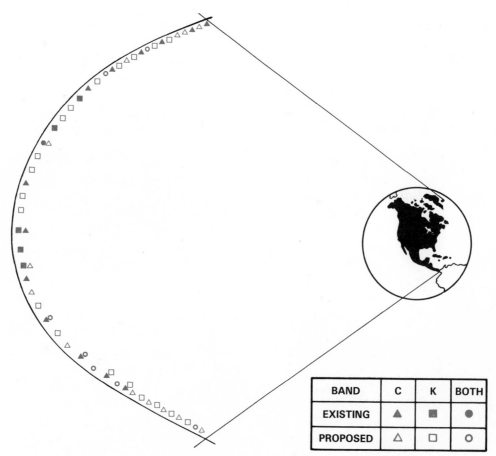

BAND	C	K	BOTH
EXISTING	▲	■	●
PROPOSED	△	□	○

Fig. 20-19. These satellite broadcasts are available in North America for the C and K bands. The C and K bands can be compared to the UHF and VHF bands on ''free'' television.

ground station in Cologne, West Germany using the same satellite.

A variety of telecommunications is done using satellites. You might be most familiar with television and radio broadcasts via (by) satellite. Perhaps your family, or a friend's family, has a satellite "dish." The dish is actually a ground station that receives signals transmitted from another point. Fig. 20-19 shows the large number of satellite "feeds" available in North America.

Satellites are also used to obtain weather information. These weather satellites send photographs of weather patterns: air currents, cloud formations, and so on. Weather satellites also carry equipment able to record temperature and humidity information. All this data helps meteorologists (weather scientists) to forecast weather patterns, Fig. 20-20.

Other users of satellites include the armed services, the government, and independent businesses. The military operates an entire fleet of satellites for communication purposes. Many large companies rely on satellites to send and receive up-to-date information.

Telecommunication through satellites is a booming industry. However, individual users are often unable to afford the cost of using satellite systems. It is thought that launching DIRECT BROADCAST SATELLITES (DBS) will open up satellite use to the individual. These particular satellites would broadcast directly to the home. Current satellite broadcasts are supplied by communications companies who own the satellites. DBS systems would have enough power to transmit signals to a smaller (and less costly) antenna than is currently needed. If DBS systems are started, satellite communication might become as common to individual users as letter writing.

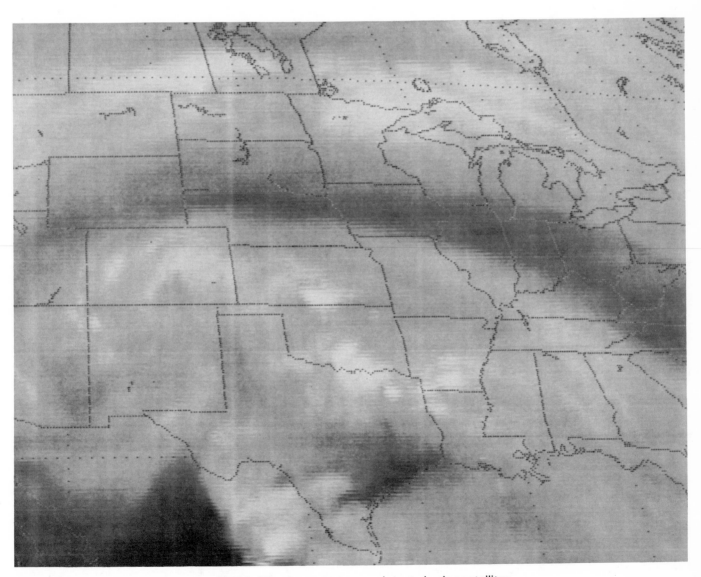

Fig. 20-20. Weather patterns are detected using satellites.

MICROWAVE LINKS

Many of the telecommunication systems discussed so far use microwaves as a link when transmitting messages, Fig. 20-21. For example, many television broadcasts travel through a network of microwave relay stations. So do many telephone conversations.

There are several reasons for the large use of microwave links. The first reason is that, with increased use, the VHF and UHF bands became too crowded. As frequencies on these bands became less and less in number, the FCC began using microwave bands.

Another reason for the large use of microwaves is that a vast amount of information can be transmitted. This is because microwaves are extremely short. They can, therefore, be concentrated (focused) into a very small, narrow beam. This allows powerful, efficient transmissions. For example, a very large and powerful transmitter can send about 100 billion bits of information per second! That is equal to about 12 television broadcasts and thousands of telephone calls.

DATA COMMUNICATIONS

Data communications is the process of transmitting information in binary form between two points. More simply, it is a process that allows computers to talk with each other. Data communication systems are used a great deal in our society. Automatic teller machines (ATM) are a good example of data communications, Fig. 20-22.

In the following sections we will discuss how data communications works as a large-scale communications model.

CODING

In Chapter 3, code was defined as a vehicle for transmitting messages. In data communication, codes are the languages computers use to communicate with one another. Most often, the languages are built into the hardware of the machine. Therefore, users of the computers seldom need to understand the coding process. If, however, two machines manufactured by different companies need to talk to each other, the user may

Fig. 20-21. Microwave towers aid in the transmission of many messages. (AT & T)

Fig. 20-22. An ATM is a smaller computer hooked up to a larger computer at a bank.

need to write some code. Writing code is commonly known as PROGRAMMING, Fig. 20-23.

Currently, two standard coding methods exist. The most common method is the American Standard Code for Information Interchange. It is commonly known as ASCII (as-key). ASCII represents 128 characters. It includes all the letters of the English alphabet, the numbers 0 through 9, and punctuation. Also included are symbols. These symbols do not represent a printed character, like a letter, comma, or number. Instead they tell the computer what to do. For example, EOT means end of transmission.

The other coding standard is Extended Binary-Coded-Decimal Interchange Code, or EBCDIC (eb-see-dik). It was designed by engineers at International Business Machines (IBM). EBCDIC is used mostly in IBM systems. It is not as widely used as ASCII.

CHANNELS

After a message has been coded, it must be transmitted. Channels, or MEDIUMS, are the physical equipment used to send the message. Basically, there are two types of channels: bounded and unbounded.

BOUNDED TRANSMISSION CHANNELS in data communications include wire pairs, coaxial cables, waveguides, and optic fibers.

Bounded transmission channels lessen noise and increase distance signals can travel. If you study a large, mainframe computer, you will see cables running from the terminals into the mainframe. These cables are an example of bounded transmission channels.

The atmosphere and outer space are two UNBOUNDED CHANNELS used in data communications. These channels are affected by noise. They need repeaters in order for signals to go a distance.

The atmosphere is used as a channel in satellite television broadcasts. The satellite codes the message and sends it down to the antenna. The antenna understands the message, then decodes it into a form we can understand.

TRANSMISSION

Data is normally sent between computers and terminals by changing the current or voltage on the channel. Transmissions are normally parallel or serial.

In PARALLEL TRANSMISSIONS, each piece (bit) of information travels on its own wire at the same time. Parallel transmission is usually used with systems located near each other. For example, a system located within one office building will usually use parallel transmission, Fig. 20-24. Wire is costly, however. So as the distance between equipment increases, so does cost. In these cases, serial transmissions are used.

SERIAL TRANSMISSIONS are used for long distance messages. The bits of information are sent one at a time over one wire. The transmission can be synchronous or asynchronous. An ASYNCHRONOUS TRANSMISSION sends each bit by itself. In SYNCHRONOUS TRANSMISSION, bits are sent in groups, or BLOCKS.

INTERFACES

An interface is a group of rules that control the way in which two machines or processes interact. These rules are usually part of the computer itself.

A modem is an interface. By changing signals over telephone wires, the modem allows one computer to talk to another computer. Or it allows

```
NO,FAIL = NO,FAIL + 1
IF NO,FAIL = 1000 THEN RETURN
IF NO,FAIL = NO,PAGE*55 THEN __
    NO,PAGE = NO,PAGE + 1 : __
    LPRINT CHR$(12)

NO,GOOD$(NO,FAIL) + CHANGED, ID$
```

Fig. 20-23. This piece of code is understood by the computer.

Fig. 20-24. All the computer needs of this company are located in one area. (Compugraphics)

a terminal in one location to talk to a central processing unit (CPU) in another location.

NETWORKS

Data communication systems are sometimes very large. In order to make them work to their best ability, they are organized into NETWORKS. Generally, there are two types of computer networks: time-sharing and distributed.

In a TIME-SHARING NETWORK any number of terminals are connected to one large central computer. Users rent part of the computer for as long as they need it. Therefore, the cost of the computer ends up being divided between all the users. This is a good method for businesses that need computers for their resources, but cannot afford to own a big system.

The data in a time-sharing network travels through telephone wires or space using radio waves. The central computer may never be seen by the user. It may be located in another city, another state, or another country.

DISTRIBUTED NETWORKS contain a central computer attached to other computers. Resources in this network are spread among all the computers. One computer does not contain all the information.

Sometimes in a distributed network, each computer is responsible for a certain task. These computers gather the information for their areas. This information can be stored in memory or sent on to the main computer.

SUMMARY

Telecommunications has many uses. A use can be a simple telephone call or a complex satellite transmission. Telecommunication systems allow us to stay in close contact with friends, relatives, and others around the globe.

Telecommunications is regulated by the Federal Communications Commission (FCC). The FCC has divided telecommunication systems into categories. These categories determine what area and frequencies a certain system can use. This aids in keeping the telecommunications industry organized.

There are many devices used to receive and transmit messages in telecommunication systems. Some devices include radio, television, telephone, teletext, videotext, satellites, and microwave links.

A data communication system is an entire system in itself. It uses its own group of devices to run the system.

KEY WORDS

All the following words have been used in this chapter. Do you know their meanings?

Amateur bands, ASCII, Asynchronous transmission, Blocks, Bounded transmission channels, Citizen band, Coding, Direct broadcast satellite, Downlink, EBCDIC, Electromechanical switching, Federal Communications Commission, Frequency, Geosynchronous orbit, Ionosphere, Mediums, Network, One-way radio, Parallel transmission, Programming, Serial transmission, Telecommunication, Teletext, Two-way radio, Unbounded channels, Uplink, and Videotext.

TEST YOUR KNOWLEDGE—Chapter 20

(Please do not write in the text. Place your answers on a separate sheet.)
1. Who regulates the use of frequencies?
2. Name one user of each FCC category.
3. The _____ is a layer of the earth's atmosphere.
4. Name two advantages of microwave signals.
5. All television broadcasts can be found in frequency bands ranging from _____ through _____.
6. Explain how images are produced on the television screen.
7. _____ _____ route telephone calls from the source to the destination.
8. Explain the difference between videotext and teletext.
9. For what does each of the following stand?
   a. CRT.
   b. UHF.
   c. DBS.
   d. ESS.
10. What is geosynchronous orbit?

ACTIVITIES

1. Send messages to friends using a variety of radios: citizen band, walkie-talkie, ham radio, etc. List the advantages and disadvantages of each system.
2. Determine the frequency of your favorite radio station. Dial this frequency into an oscilloscope to actually "see" the wave.
3. Develop a technical illustration of how a geosynchronous satellite provides worldwide service.
4. Design a telephone system for use in your classroom. Use students (instead of electronic devices) as relays, switches, and telephone equipment.

Satellites can be used to send telephone messages around the world. (AT & T)

21 | Light Communication

The information given in this chapter will enable you to:

○ *Describe the purposes and uses of light in modern communication systems.*

○ *Understand how light signals are created and received by humans and machines.*

○ *Explore the uses of laser and fiber optic technologies.*

○ *Develop and transmit messages with beams of light.*

Although it may seem strange, light is a very common communication medium. Can you think of some typical examples? Lighthouses warn ships of dangerous shorelines or shallow waters. Traffic lights control the flow of cars and trucks on roadways. Neon signs and lighted displays are effective for advertising in stores. The projectors used in your school classroom are a form of light communication. The latest compact disc players rely on laser light for producing clear sounds and pictures. This chapter will explore these and many other uses of light in modern communication.

PURPOSES AND USES

The purpose of communication technology is to help transmit information, Fig. 21-1. Light communication is no exception. It is useful in many control and information systems. The following sections explain modern and future uses of light.

TRAFFIC CONTROL

Many forms of transportation rely on signs and signals to direct their flow. You can probably give several examples for directing cars, trucks, etc. However, the direction of sea and air traffic is also critical, Fig. 21-2.

Light signals are very important in ground travel. Lights at busy intersections direct cars and trucks. Flashing lights along roadways inform us of hazardous areas. In addition, various signals inform railroad engineers if any trains are on the tracks ahead. This is known as the BLOCK SIGNALING SYSTEM, Fig. 21-3.

Many forms of light are used in directing traffic in the air, too. Have you ever seen a rotating airport beacon at night? The green, red, or white signals guide airplanes as they fly overhead. Other lights line the runways to assist approaching aircraft with their landing. Flashing strobe lights also "point" the direction for planes headed toward runways, Fig. 21-4. Light communication is very important to the safe operation of any airport system.

Fig. 21-1. Many presentations and oral reports are aided by showing vital information on a screen. This speaker is using an overhead projector and transparency to improve his talk. (Bell & Howell)

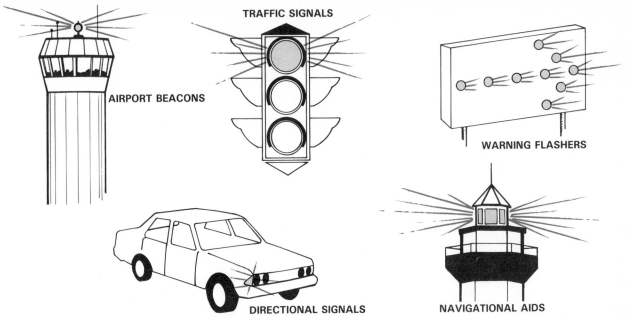

AIRPORT BEACONS

TRAFFIC SIGNALS

WARNING FLASHERS

DIRECTIONAL SIGNALS

NAVIGATIONAL AIDS

Fig. 21-2. Here are examples of light communication used in regulating traffic in the transportation industry.

Fig. 21-3. A block signaling system alerts railroad engineers of conditions on the tracks ahead. (Sante Fe Railway)

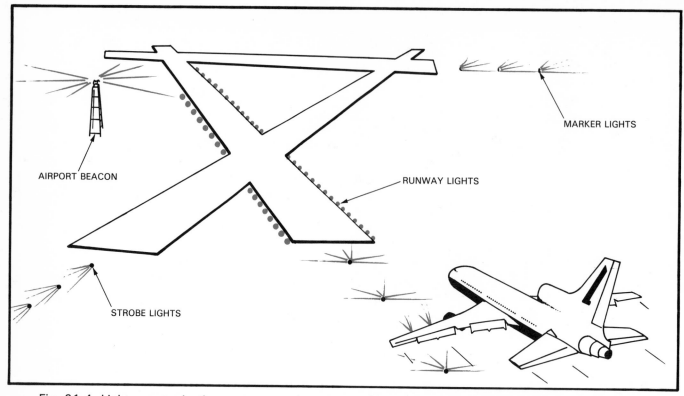

Fig. 21-4. Light communication systems at airports are critical to the safe landing and take-off of aircraft.

Water, or MARINE, vehicles also rely on light signals for direction and safety. Perhaps the best example of this type of signal is a lighthouse, Fig. 21-5. The earliest "lighthouses" were simply fires built on high places. Today lighthouses operate electrically and have strong signal beacons. During the day, most lighthouses can be recognized by their shape or distinctive stripes.

Many other forms of light communication are used to guide boats and ships. BUOYS mark channels (deeper water) and harbors. In recent years, light towers have been built miles off coastlines, in the water. These large structures serve as points of reference for mariners. Their bright lights mark the entrance to local ports, Fig. 21-6.

ADVERTISING

Another important use of light communication is in advertising. Signs, banners, and other visual displays seem to be everywhere in our communities. Most are lighted to attract customers attention, Fig. 21-7. These lighted displays identify shops and create interest.

Can you think of ways signs are lighted? Many signs are merely boxes with fluorescent lights inside. The plastic or glass coverings create an appearance of shape or color. Other advertising displays have flashing border lights. Colored light

Fig. 21-5. Among the earliest uses of light communication was signaling ships about dangerous waters or coastlines. Lighthouses, like this one, line the coast of our country.

Fig. 21-6. This large light tower is positioned several miles offshore to mark the entrance to a harbor. The rotating beacon can be seen for miles.

bulbs draw attention to the message on a sign. A common type of colored light is produced using neon gas. These are called NEON LIGHTS.

Neon lights are made by glass blowers, Fig. 21-8. Glass tubing is heated and bent into the shape of letters and designs. The tubing is then filled with neon gas. The gas acts as a conductor of electrical current. As electricity flows through the tube, the gas glows brightly. Different colors are produced by mixing various types of gas with the neon.

Flashing lights are another technique used to create interest in advertisements or other displays. Flashing lights usually attract more attention than static light displays. They create movement and repetition. Examples of flashing lights include portable roadway signs and signs for movie theaters. Warning signs usually have flashing lights, too. Red or yellow flashing lights signal dangerous areas, Fig. 21-9.

Flashing lights can work in two ways. In one method all the lights go on and off at the same time. This creates the flashing affect but no movement. The second way is to have an electric current turn each bulb on in sequence. When one light goes off, the next one comes on. This procedure repeats itself and movement is created. This is often seen on theater marquees (signs).

An additional use of light in communication is evening ILLUMINATION. Lighted signs and displays line our streets and roads. Billboards are

Fig. 21-7. These advertising displays often attract our attention as we pass them on the roadway.

Fig. 21-8. Neon lights are popular for commercial signs and advertising displays. This sign shines brightly 24 hours a day to inform others of the firm.

Fig. 21-9. The red flashing lights on this post warn individuals to stay clear of the area due to high voltage.

often equipped with bright lights to help travelers to read them at night. Public buildings and monuments are also illuminated at night.

As these examples have shown, light is very important to advertisers and related businesses. It brings life to many visual displays. What would your community be like without lighted signs and billboards?

INFORMATION TRANSMISSION

Light has always been important in the transmission of information. The earliest means of transmitting light signals was by REFLECTION. Mirrors and other shiny objects were used to reflect the sun's rays. Pioneers heading west in frontier days signaled other travelers by this same method. The U.S. Navy still sends messages between ships using a series of light signals.

These examples of light communication involve the use of basic technical know-how. But we are also informed by modern light systems. In airports, lighted signs mark gates and exits, Fig. 21-10. Radio and TV stations use lighted signs to warn others of "live" shows being broadcasted. Different types of lights in elevators show us which floor we are on while in skyscrapers.

Indicator lights are found on almost all of our technological gadgets. Lights show us if our ovens, freezers, and microwaves are working. Display lights on the dashboard of a car indicate trouble. The same holds true for industrial machinery. Various lights inform us if each machine is functioning properly, Fig. 21-11.

Many machines exchange messages or "read" information by light signals. One example is a computer. A light pen is often used to enter information into a computer system, Fig. 21-12. In the construction and automobile industries, laser light is useful in determining distances. Laser light acts as a surveying tool in measuring land areas. In the design of new cars, laser light helps determine sizes and shapes, Fig. 21-13.

Modern information systems also use light as a communication medium. You learned in Chapter

Fig. 21-10. This attractive display sign informs travelers the direction to various parts of the airport terminal.

Fig. 21-11. This production worker monitors the function of a machine by watching indicator lights. If trouble develops along the line, warning lights will flash on the panel in front of him. (Hammermill Paper Co.)

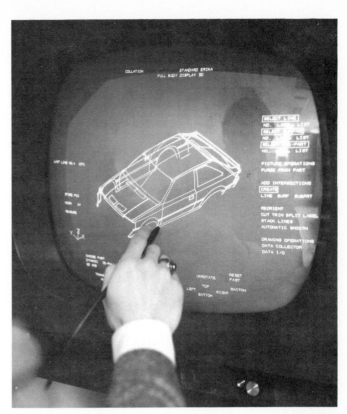

Fig. 21-12. This designer is "drawing" on the computer screen with the aid of a light pen. With this computer, the lines of the sketch can be altered easily and quickly. (Ford Motor Company)

Fig. 21-13. A thin beam of laser light is often used to measure and record data in various industries. The shape and size of this clay model are being read by a laser. A computer can then prepare drawings of the model. (Chrysler Corp.)

19 that fiber optics are coded messages transmitted as pulses of light. Light, in the form of lasers, is also used to record vast amounts of data on disks, Fig. 12-14. Businesses keep their vital information on small disks rather than in large books. At home, compact disc players are becoming popular for stereo and video systems. Information in the form of audio sounds and video images is read by small lasers.

ENTERTAINMENT

Light is also used for entertainment purposes. Both slide and movie projectors use focused light to project images on a screen. Perhaps the most common example of PROJECTED IMAGERY is a movie shown at a theater.

Another exciting use for light is in concerts and stage shows. Large spotlights are used to light the stage area, Fig. 21-15. Brightly colored lights add much to the musical entertainment. In fact, light displays are often just as "exciting" as the music at many concerts and shows.

In museums, special three dimensional (3-D) images are created with light. These displays are

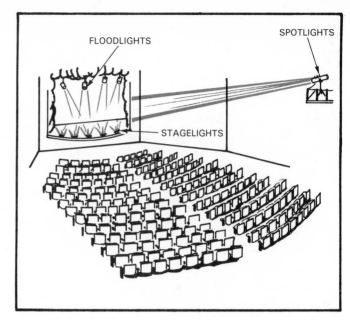

Fig. 21-15. Theater lights illuminate the stage area from many angles.

known as HOLOGRAPHS. Most holographs are developed by beams of light intersecting in space. A lifelike image is produced at the point where the light waves meet. Holographs may one day have important commercial value far beyond the recreational uses of the present.

As has been shown, light is used in many ways. As a communication tool, visible light has become quite popular. Another form of light, the laser, is presented in the next section of this chapter. We will explore the usefulness of lasers in modern communication.

LASERS IN COMMUNICATION SYSTEMS

The term LASER is an acronym. An acronym is a word made up of the first letters of a series of words. The acronym laser comes from the words Light Amplification by Stimulated Emission of Radiation. Therefore, laser light is a form of radiation that is amplified (boosted) to a high energy level. Laser light produces a strong, narrow beam of light. It does not disperse (spread out) and become lost energy when used. It can be focused on a very tiny point, a long distance away.

Laser light is COHERENT energy. This means that the length of the waves that make up the light are all the same frequency. In the case of laser light, the wavelengths are quite short. Because they are so short, they are ideal for communication systems. With short wavelengths, messages can be sent faster and in large groups. A single pulse of light can contain thousands of messages. In addi-

Fig. 21-14. Over 100 million bits of data can be recorded on this optic laser disc. High speed methods of communicating with lasers are greatly improving our ability to store and retrieve information. (RCA)

tion, these signals can travel along the same medium at the same time. Laser light signals can be sent through the atmosphere or through transparent materials, like glass.

Sending lightwaves through the atmosphere has major drawbacks. For example, beams of light only travel in straight lines. This is often called LINE OF SIGHT transmission. Light energy cannot travel around corners of buildings or over hilly terrain. Another drawback is that these signals are easily interrupted. Light signals are often affected by poor weather conditions, such as fog, rain, or clouds. Interference among the individual light signals also creates problems.

In recent years, a new system of transmitting light signals has become quite popular. It avoids the drawbacks of sending signals through the atmosphere. This method of communication is known as FIBER OPTICS. These signals are sent as pulses of light along a transparent (clear) fiber, Fig. 21-16. With fiber optics, there is very little lost energy.

Optical fiber cables currently in use are made of glass, plastic, or a mixture of silica and other compounds. It is important that fiber optic cables be free of impurities (defects and flaws). Flaws and defects interfere with the light signals being sent. Fewer impurities means better transmissions.

Fiber optic technology may someday replace traditional electrical wires and circuits. For that reason, we should explore this technology in more depth. The following sections explain how fiber optic communication systems function.

TRANSMISSION OF FIBER OPTIC SIGNALS

Many telephone and cable TV companies use fiber optic cables for sending messages, Fig. 21-17. At the source of the transmission, signals (sound

Fig. 21-16. In fiber optics, pulses of light travel along transparent lines to transmit messages and data. Several of these transparent lines can relay millions of telephone calls at the same time. (Fibronics International)

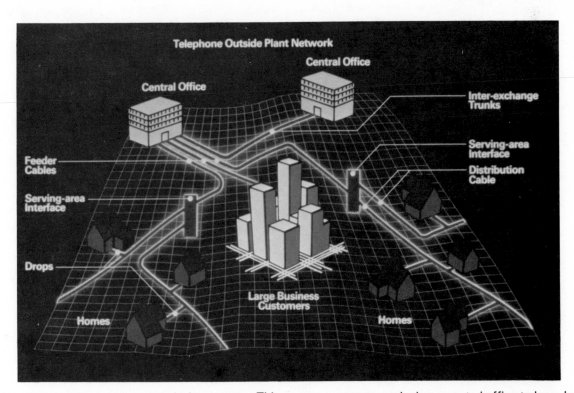

Fig. 21-17. A typical fiber optic transmission system. This system connects a telephone central office to large businesses and neighborhoods. (Corning Glass Works)

or video images) must be changed into pulses of light. This light source can be one of two forms: light emitting diodes (LEDs) or lasers.

An LED gives off visible light. A common use of LEDs is on calculator displays. When used in fiber optics, LEDs are less costly and longer lasting then lasers.

A laser emits (gives off) infrared light. Despite costs, lasers are used more often than LEDs in fiber optics. This is because lasers are more powerful. They have the power to send a very strong signal.

Optical fibers have a number of advantages over standard copper cables. For example, optical fibers can carry a much greater number of signals than a copper cable. This is because light travels faster than electrical current.

Another advantage is that optical fibers are many times smaller and lighter than copper cables, Fig. 21-18. Optical fibers are about the thickness of a human hair. They can be used in places where there is little space. And they are free of most of the noise that occurs with copper cables.

A third advantage is that light signals do not fade as quickly as electrical signals, Fig. 21-19. Copper cables need repeaters about every mile to

Fig. 21-19. Cross-section of an optical fiber. It has been enlarged many times to show the internal structure of the line. Light flows through the center of the wire. Any light trying to "escape" is reflected back toward the center of the line. (Fibronics International)

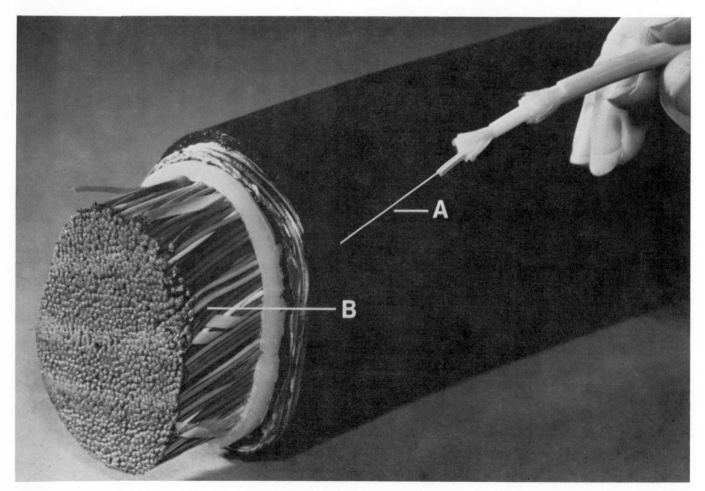

Fig. 21-18. Compare these cables. A—Typical fiber optic wire. It is actually the thickness of a human hair. B—Copper wire. It carries the same amount of information as the fiber optic wire. (Corning Glass Works)

boost signals. Optical fibers, however, only need repeaters about every 14 miles.

When a signal reaches its destination, it must be decoded. In this case, the light signal must be changed into an electronic signal. For example, if the destination is a computer, the signals must be decoded into something that the computer will understand.

SUMMARY

Light is a very common communication medium. It can be as simple as a traffic light or as complex as a fiber optic transmission. Light communication is used often in the traffic control, advertising, information transmission, and entertainment industries.

An important "high tech" use of light for communication is the laser. It can be used to communicate with spacecraft and satellites. Lasers are also used in the fiber optics industry. Along with LEDs, they are used as a light source for sending thousands of messages over thin, transparent wires.

Light communication is a growing technology. When advances are made in light communication, they are often made in other technologies as well.

KEY WORDS

All of the following words have been used in this chapter. Do you know their meanings?

Block signaling system, Buoys, Coherent light, Fiber optics, Holographs, Lighthouse, Line of sight transmission, Illumination, Laser, LED, Neon lights, Reflection, and Traffic control.

TEST YOUR KNOWLEDGE—Chapter 21

(Please do not write in this text. Place your answers on a separate sheet.)
1. List three examples of light communication used in traffic control, entertainment, and advertising.
2. The _____ _____ _____ uses light signals to tell railroad engineers if any trains are on the track ahead of them.

MATCHING QUESTIONS: Match the definition in the left-hand column with the correct term in the right-hand column.

3. __ Used in the evening to make signs and streets more visible.
4. __ Three-dimensional image created with light.
5. __ Used to mark channels of deep water and harbors.
6. __ The earliest method used for transmitting light signals.
7. __ Colored light made by glass blowers.

a. Buoys.
b. Holographs.
c. Neon lights.
d. Illumination.
e. Reflection.

8. What does the acronym, laser, mean?
9. Name two disadvantages to sending laser transmissions through the atmosphere.
10. In fiber optics, signals are sent as _____ of light along thin, clear fiber.

ACTIVITIES

1. Waves of light from the sun are INCOHERENT. This means the waves of light are of different lengths (frequencies). Split a beam of sunlight with a prism to see these different lengths. Each color is a different frequency. Identify which colors have the highest frequencies.
2. Make a list of light communication devices located in your home and in your school. Create a display board showing each of these devices.
3. Invite a representative from a telephone company to talk to your class about the latest uses for fiber optic systems. Ask the representative to explain how the system works, what size load it can handle, and how much it costs to install and maintain.
4. Research the use of Morse code. Learn several letters and numbers, in order to send a message to a classmate. Use a flashlight to send the message.

22 | Acoustical Communication

The information given in this chapter will enable you to:
- *Recognize how audible signals are transmitted.*
- *Name several acoustical communication systems.*
- *List the components of a radio/stereo system.*
- *Explain the processes for making a record and a tape.*
- *Define radar terminology.*
- *Discuss sonar operation.*

Acoustical communication involves transmitting information with sound, both audible and inaudible to humans, Fig. 22-1. You will recall that AUDIBLE sounds are those that can be heard by humans. INAUDIBLE sounds are also sounds, even though they cannot be heard by humans. A dog, for example, can hear many sounds that are inaudible to human ears.

HOW SOUND WAVES TRAVEL

Like radio signals, sound also travels in waves. But these waves are created when there is a disturbance in a medium that has mass and is elastic.

Think of a wave created by your hand in a swimming pool, full of water. Your hand creates the disturbance which starts the wave. The medium is the water. The MASS of this first wave is the quantity (amount) of matter that forms the wave. As this wave travels through the water, it builds up strength and speed. This is called MOMENTUM. The mass of the first wave has, therefore, built up momentum. It will continue until it hits the side of the pool or fades due to lack of momentum.

The wave also has elasticity. ELASTICITY is the ability of a mass to return to its original shape after an outside force has been removed. This means that after one wave travels through the

water, the water returns to its original shape. Then another wave can travel through the water. This motion will continue until the source of the motion (your hand) stops.

It is very important to note then that sound

Fig. 22-1. This musician is creating sound with his voice and his guitar. (Ball State University)

waves do not cause the medium to move. However, they are carried through the medium if it (the medium) has mass and is elastic.

Let us use an example to put all this information together, Fig. 22-2. Suppose someone standing next to you claps his or her hands. The pressure of two hands coming together (mass) creates a wave in the air surrounding you. The elasticity of the air allows the wave to ripple through the air. The wave then passes through your ear canal and past your eardrum. The eardrum also is elastic. It vibrates from the pressure created by the wave. The frequency and pressure of the sound wave is duplicated (copied exactly) by the eardrum.

In acoustical communication, messages can be transmitted by sound waves (acoustical energy). The distance the message will travel depends on pitch and loudness. These make up the power of sound.

PITCH is determined by the frequency of the wave. A high frequency wave will create a high-pitched sound. LOUDNESS is determined by the pressure (strength) of the wave. Therefore, a high-pitched, loud sound will likely travel farther than a low-pitched, soft sound. In addition, all sound waves are affected by interference from objects or natural occurrences (rain, snow, wind). This will also affect the distance a message will travel.

Acoustical information can also be transmitted using electrical energy. With this type of transmission, the signal receives help in covering a longer distance. The help comes in the form of electricity. Radio and the telephone are just two examples of acoustical information being transmitted by electrical energy. A receiver or speaker converts the electrical signal to a sound we understand.

In acoustical messages transmitted by the telephone, elasticity comes in the form of a diaphragm. A DIAPHRAGM is made up of a flexible membrane (thin medium) attached to a solid frame, Fig. 22-3. When sound passes the diaphragm, it vibrates in the exact way the original sound is vibrating. When electric current is added, the message can be carried over wires for long distances. The invention of the diaphragm led directly to the invention of the telephone.

USES AND PURPOSES

Acoustical communication is used in many ways and for many purposes. Perhaps the most common use is for entertainment. Records, tapes, and compact discs of popular music are all examples of entertainment. A sonar unit uses acoustical waves for measuring and locating purposes. It is also used to aid the handicapped. For example, to aid the blind, books are often read aloud and recorded on tape, Fig. 22-4.

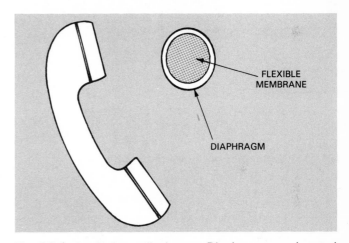

Fig. 22-3. A telephone diaphragm. Diaphragms are located in the receiver and in the transmitter of the handset.

FLEXIBLE MEMBRANE

DIAPHRAGM

Fig. 22-4. Audio tapes of books enable blind people to enjoy written works. (Recording for the Blind)

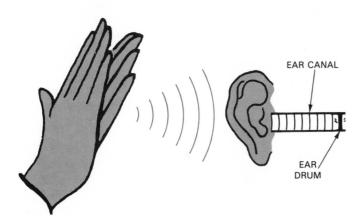

Fig. 22-2. Can you identify the source, medium, and destination in this drawing?

EAR CANAL

EAR DRUM

ACOUSTICAL SYSTEMS

Bells and horns can communicate many different messages. An alarm in the morning lets you know when it is time to wake up. The siren from a fire engine warns of a dangerous situation. The sound of a car horn might also mean danger.

Public address (PA) systems and loudspeakers can be found in many places. Nearly all schools and offices have PA systems. At school, a PA system might be used to make general announcements. In an office setting, it might be used to page (call) employees.

Another acoustical communication system we all may know is the stereo system. A stereo system can be used for entertainment or information purposes. This system will be discussed later in the chapter.

Compact discs are another method for recording sounds, Fig. 22-5. However, rather than a needle "reading" the groove in the record, a laser beam "reads" the disc. The sound itself is recorded as a series of pits on the surface of the disc. When the disc is played, it rotates past the laser. Changes in the surface of the disc cause the laser signal to fluctuate (change). This signal produces a digital

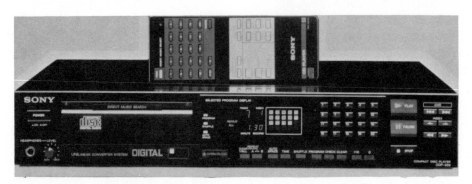

Fig. 22-5. Compact disc player. (Sony Corp.)

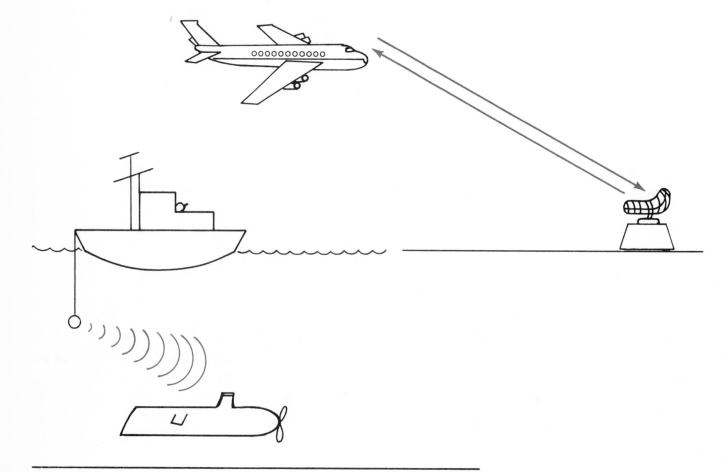

Fig. 22-6. Radar and sonar systems are used largely by government and industry.

signal, which is then changed back into sound.

Compact discs have many advantages over records. One of the most important advantages is the quality of the sound. Dust and wear on a record or tape can be read by the needle or tape head. This becomes noise in the sound transmission. But with compact discs, the laser does not read dust. It passes over it. In addition, there is no needle to wear away the groove. Therefore, compact discs avoid the problem of sounding "old" and worn out. As the cost of compact discs and disc players decreases, this type of acoustical system will become more popular.

You may not be very familiar with radar or sonar systems, Fig. 22-6. RADAR uses inaudible sound waves to detect solid objects in the path of the waves. In SONAR, sound waves are used to send signals through water. Sonar is rarely used for communication between two points. Rather, it is used for such things as measuring water depth and detecting the presence of solid objects. These objects may be either on top or underneath the water. These two systems will also be discussed later in the chapter.

MICROWAVE LANDING SYSTEMS (MLS) aid aircraft in landing. A beam is transmitted from the MLS antenna. This beam scans the runway area up and down, and back and forth, Fig. 22-7. When an airplane enters this area, the microwaves are blocked. This signal is reflected back to a computer in the control tower. This information will give the position of the airplane, and aid the controller in directing the aircraft through a smooth landing. Guidance of this type is more precise than radio control. This makes flying safer.

RADIO/STEREO SYSTEMS

Do you have a large collection of records and tapes? Do you know someone else who does? The recording industry is large and profitable. Every year, millions of music enthusiasts (those who enjoy music) buy millions of recordings. In order for them to enjoy their records or tapes, stereo systems are required.

The stereo system is designed to give a full, lifelike transmission of a recording. Stereo systems come in many forms. They can be small and portable, like a personal cassette player. Others can be large and complex. Power can range from weak to strong.

No matter what the size or power of a stereo system, however, they usually contain the following equipment: receiver, amplifier, playback unit (turntable/tape deck/CD player), and speakers, Fig. 22-8. These components contribute to the overall sound quality produced by the system. One

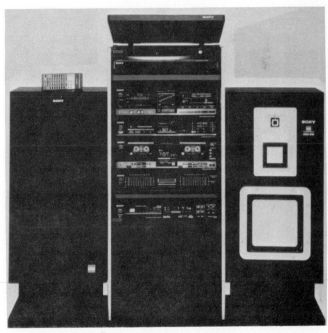

Fig. 22-8. A variety of stereo systems are available to consumers. This system is very complex. (Sony Corp.)

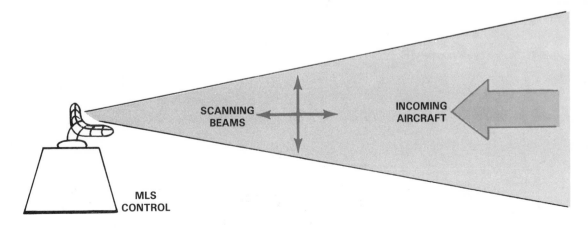

Fig. 22-7. At airports, MLS are important acoustical communication devices.

bad piece of equipment will cause the entire system to sound bad.

RECEIVER

The receiver in a stereo system tunes in FM and AM radio signals. This is the component that contains the buttons for selecting the bands (AM, FM) and for selecting the radio station. The receiver is also known as the TUNER, Fig. 22-9. In simple systems, it is called a radio.

AMPLIFIER

The amplifier is the most important part of a stereo system, Fig. 22-10. It is the amplifier that accepts electrical signals from the turntable, tape deck, or radio receiver. The amplifier then powers (boosts) the signals enough to activate the speakers.

A good amplifier will closely reproduce the sound. That is, the final sounds are nearly identical to the transmitted signal. In addition, it will boost the signals without introducing interference of any kind.

SPEAKERS

Speakers change electrical energy back into acoustical energy. They should reproduce a broad range of sounds. Just how "good" a speaker sounds, however, depends largely on personal taste.

Speakers come in several types. CRYSTAL speakers and ELECTROSTATIC speakers are often used along with other speakers for quality reproduction.

A DYNAMIC speaker works using a wire wrapped around a permanent magnet. The magnet is mounted next to a diaphragm. When the electric current comes through the wire, a magnetic field is created between the magnet and the diaphragm. As the magnetic field changes, the diaphragm vibrates. This reproduces the sound.

A speaker produces sound by making sound waves in a cone-shaped area. Each speaker has two of these areas. They are called the WOOFER and the TWEETER, Fig. 22-11.

Low-frequency sounds are sent out through the woofer. This is because low-frequency waves move slowly back and forth over a large area. The

Fig. 22-9. A typical receiver or tuner. (Sony Corp.)

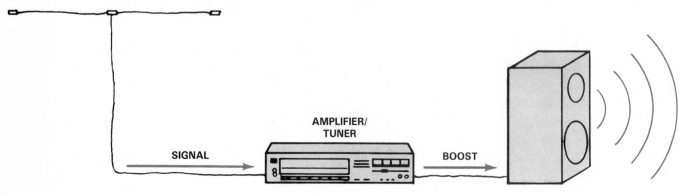

Fig. 22-10. Amplifiers are often housed along with the tuner in modern stereos.

woofer is large enough to send out a low-frequency wave with good sound.

The tweeter is used to send out high-frequency sound. High-frequency waves move rapidly over a small area. Therefore, the smaller tweeter is used to produce high-frequency tones.

The sound reproduced by speakers depends a great deal on where they are placed. For example, in a room with a low ceiling that contains a great deal of furniture, volume may need to be turned up. This is because the sound is absorbed by all the furniture. Just the opposite is true for a room with high ceilings and a few pieces of furniture, Fig. 22-12.

TURNTABLE

The turntable is used to spin records. This must be done at a constant (never changing) speed. Constant speed insures the proper sound. If the turntable spins too quickly or slowly, the sound will waver and become distorted.

A second important function of the turntable is to produce an electrical signal from the grooves of the record, into electrical energy. This is done by the tonearm. The tonearm is located next to the turntable platter. The TONEARM contains the cartridge and the stylus, Fig. 22-13. The STYLUS actually picks up the signals from the grooves. It is a very small, flexible metal strip. At the end is a hard material, usually a diamond. This is often called the NEEDLE.

The CARTRIDGE accepts the vibrations picked up by the stylus. It changes these vibrations into electrical energy. The energy is then sent through a wire to the amplifier.

TAPE DECK

The tape deck contains a tape recorder. It is used to record and/or play audio tapes. The recording or playing of a tape is done by STATIONARY

Fig. 22-12. Top. Sound can travel much farther in a room with a few pieces of furniture and a high ceiling. Bottom. The furniture and plants in this room absorb a great deal of sound.

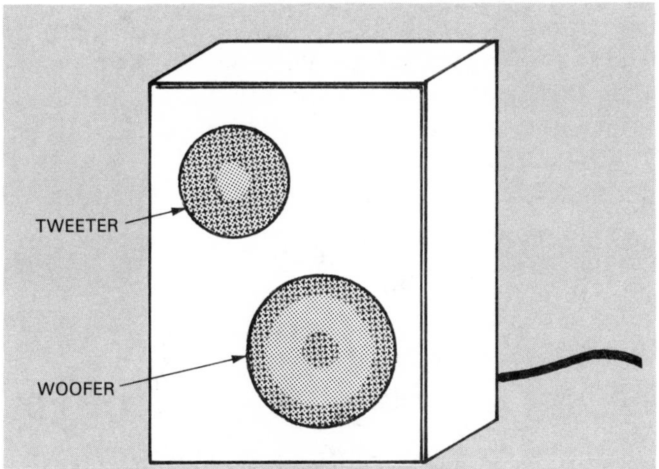

Fig. 22-11. Stereo speakers are an important part of a stereo system. An expensive stereo only sounds good if it is attached to quality speakers.

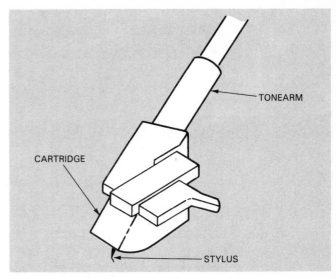

Fig. 22-13. Note the locations of the various parts on this tonearm.

HEADS. Stationary means they do not move. Good machines have three heads—one for erasing, one for recording, and one for playing.

There are three types of tape recorders: cassette, open-reel, and cartridge. You are probably familiar with CASSETTE RECORDERS. They use cassette tapes. When the cassette is inserted into the recorder, it passes by the playback head.

An open-reel recorder uses open-reel tape. The tape starts out entirely on one reel. It is threaded past the heads to the TAKE-UP REEL, Fig. 22-14. When all the tape is used, it will all be on the take-up reel. An open-reel recorder is also known as a reel-to-reel tape recorder.

A cartridge recorder also uses tape that can be put in the machine without threading. However, it contains one loop of tape. Rewinding is not needed with cartridge tapes. An 8-track tape is a familiar type of cartridge tape.

All of the equipment discussed is important to the entire stereo system. It allows the use of recorded acoustical communication.

RECORDS AND TAPES

How useful would a stereo system or tape recorder be if there was nothing available to play on it? Records and tapes are used to store and retrieve (play back) acoustical information.

RECORDS are vinyl discs. They contain a series of grooves. These grooves represent waves of pressure. The records we buy today are usually STEREOPHONIC RECORDINGS. Stereophonic recordings are made using several microphones. They are placed in different areas of the recording studio. MICROPHONES are used to change human voices and musical instruments into electrical signals. As electrical signals, they can be recorded.

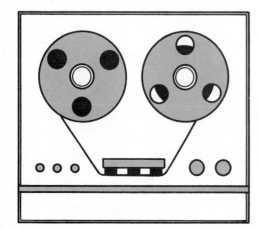

Fig. 22-14. An open-reel recorder is used often in audio production work.

Recording or STORING acoustical information is done in the following way. The vinyl record has a V-shaped groove. As information is being recorded, a cutter head moves up and down in the groove. The cutter head is controlled by two rods. They are attached to either side of the cutter head. Each of these rods is connected to one of the microphones. When information from one microphone is being recorded, the matching rod vibrates the cutter head. This causes the cutter head to cut a TRACK on the wall of the groove. Information from the other microphone cuts a track into the opposite wall, Fig. 22-15. Therefore, stereo recordings have two tracks.

TAPES are plastic strips coated with a magnetic material. When a recording is made on tape, sound waves from the microphone are changed into electrical signals. These signals are sent into an electric coil. This coil is known as a RECORDING HEAD. As the tape passes by the head, the electrical signals change the coating of the magnetic material on the tape. Refer to Fig. 22-16.

OPERATIONS

Playing back, or retrieving sound recordings is a simple process. Tape recordings are played back by running the tape past a PLAYBACK HEAD. It must pass the playback head at the same speed at which it was recorded. The electrical signals on the tape create an electrical current in the coil on the head. This current will reproduce (copy) the sound recorded. The signal is amplified and projected through a speaker system, Fig. 22-17.

The stylus is the part of the turntable that "reads" the information from a record. The stylus follows the V-groove, reading both tracks. Each track is heard at the same time through separate speakers. As we learned earlier, the stylus vibrates according to the tracks. The vibrations are sent to the cartridge. Here they are changed to electrical signals and sent through to the amplifier.

These recording methods and playback procedures may soon change, however. Digital hi-fi systems are gaining in popularity. They allow for better sound quality. We will look at digital hi-fi in the next section.

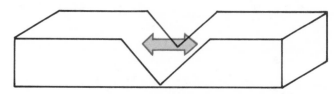

Fig. 22-15. Cutting acoustical information into vinyl.

DIGITAL HI-FI

Have you listened lately to a compact disc recording? Did you notice the quality of the sound? Did the music sound "live?" Compact disc players are a component of the digital hi-fi system.

Standard recordings store acoustical information in analog form. You will recall that ANALOG means continuous. The tracks or magnetic particles are exact copies of the sound wave created by the voice or instrument.

With digital recordings, the analog waves are changed into digital code. When the recordings are played on digital equipment, the code is changed back to analog waves to reproduce the original sound waves. This process works much the same way telephone signals are converted in digital phone systems.

Earlier in this chapter, we discussed compact discs. The advantages pointed out are also true for digital hi-fi systems. They are all part of the same family. And like compact discs, they become less costly every day. As the system is perfected, it will become as common as the traditional stereo system.

RADAR AND SONAR SYSTEMS

We are all familiar with weather forecasts. The weather maps shown on TV are made using radar. RADAR stands for RAdio Direction And Ranging. It is a radio-based device that is used to detect objects that are too far away to be seen with the eye. Radar is often considered acoustical communication because it uses reflected radio waves. Radio and sound waves are a form of electromagnetic energy.

SONAR functions much the same as radar. Sonar, however, travels through water instead of the atmosphere. It is a system that uses reflected waves to detect underwater objects or ocean depths. Sonar can detect old sunken ships or schools of fish, to name a few. Sonar stands for SOund NAvigation and Ranging.

RADAR OPERATION

Radar uses microwaves to detect distant objects. The microwave energy is sent out in short bursts from a transmitter and antenna, Fig. 22-18. The microwave is focused into a narrow beam. If a

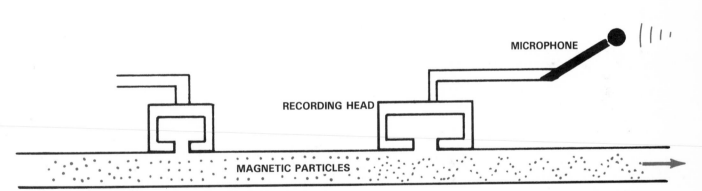

Fig. 22-16. Making a recording on tape. Notice how the tape differs after passing the recording head.

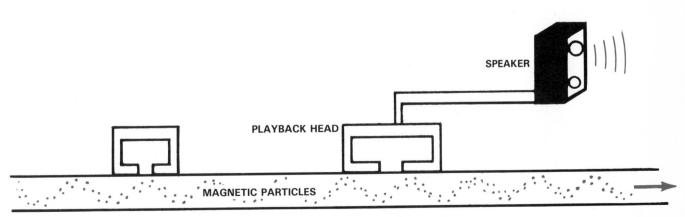

Fig. 22-17. Playing back a tape recording.

solid object is in the path of that beam, some of the microwave energy will be reflected or ECHOED back to the antenna. This is known as DETECTION. The antenna sends it back to a receiver. The receiver changes the echo into electronic signals. The signals show up as spots of light on a display screen. The spots are called BLIPS.

Types of Radar

There are two types of radar: pulse radar and continuous wave radar. PULSE RADAR sends out timed bursts of energy. The distance and direction of an object are measured by these pulses.

Aircraft carry pulse type radar systems for guidance and safety. The pulses are sent out several hundred times per second. The beam is sent straight ahead and to both sides. This gathers information in an arc in front of the plane.

Airport radar is also pulse type. However, it sends out more powerful pulses than airplane radar. Also, it gathers information in all directions.

CONTINUOUS WAVE RADAR sends out a constant flow of energy. It is used to measure speed of objects. An object that is still will echo the same frequency as what is being sent. A moving object, however, will send back a different frequency. Speed of the moving object is measured in relation to the stationary (still) radar.

Radar used by police to detect speeding cars is continuous wave radar. Continuous wave radar makes use of the DOPPLER EFFECT. The Doppler Effect explains the change in frequency of a sound source as it approaches and then passes another object, Fig. 22-19. For example, as a car approaches and then passes a radar source, there will be a change in the frequency of the sound waves reaching the radar. The change is caused by the movement of the car. Based on this information, the radar determines the speed of the car.

SONAR OPERATION

Sonar works using sound waves instead of microwaves, Fig. 22-20. Since sound is actually a wave of pressure, it moves easily through water. In contrast, microwaves are quickly absorbed by water. As we learned earlier, sonar is often used to learn water depth. It is also used by ships to detect submarines and by submarines to detect ships. Sonar is not especially useful for person-to-person communication.

Sonar uses echo location to detect objects. A pulse of acoustical energy is transmitted through the water. This sound wave travels until it strikes an object large enough to reflect the original wave. The reflection of the original wave travels back

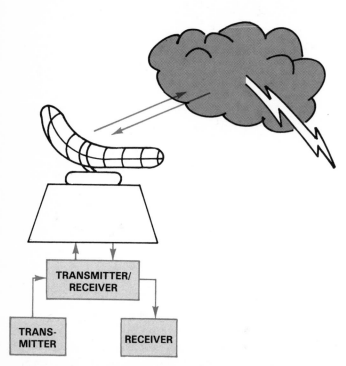

Fig. 22-18. Radar waves detect an approaching storm. The signals are processed and show up on the display screen as a blip.

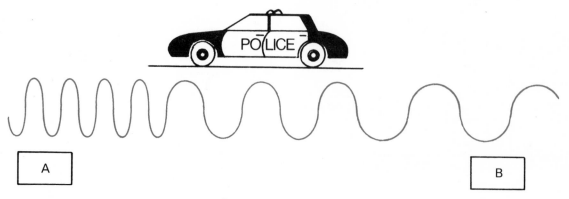

Fig. 22-19. With the Doppler Effect, the frequency from Point A to Point B will vary.

toward the source of the sound. This reflection is detected as an ECHO at the source. The size and distance of the reflected object can be learned by studying the echo. The amount of original wave that was reflected will give the shape. The time that passed between the original transmission and the receiving of the echo determines the distance.

Sound waves in sonar are made by a TRANSDUCER. The transducer is a device that acts like an underwater microphone and loudspeaker. It both picks up and sends out sound waves.

Types of Sonar

Two types of sonar are used commonly: passive and active. PASSIVE SONAR is used to listen for sounds. It does not transmit sound waves. With the use of passive sonar, position, distance, and depth cannot be learned.

ACTIVE SONAR sends out and listens for sound waves. It is used when distance or depth must be known.

Basically, there are three ways to use active sonar. It can be used by ships by dropping a transducer, much like an anchor, Fig. 22-21. Any

echoes sent back reach a receiver on the bottom of the ship.

A second method for using active sonar is the SONUBUOY, Fig. 22-22. This is a transducer with a collar to keep it afloat. Sonubuoys float in the

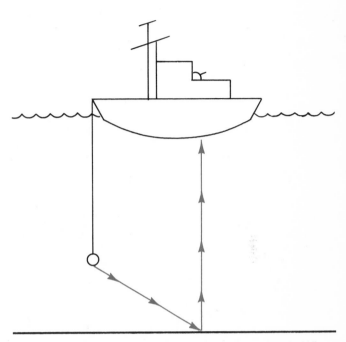

Fig. 22-21. This ship uses sonar to detect sea depth. When the signal detects bottom, an echo is reflected back to the ship.

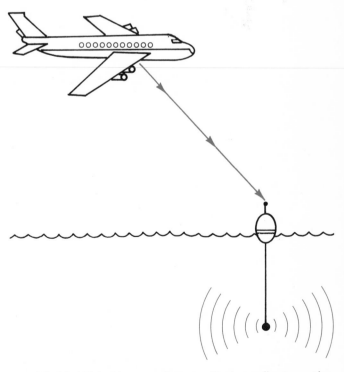

Fig. 22-20. This U.S. Navy sonar is used in search and recovery work. The sonar creates a visual image of the sea floor. The image appears on the ship's monitor. (U.S. Navy)

Fig. 22-22. With this type of sonar, the buoy floats on the water. It is attached to a transducer, which is located below the water.

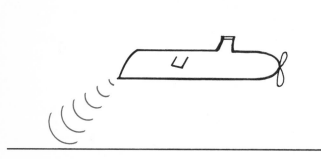

Fig. 22-23. Fixed sonar being used on a submarine.

water, with the transducer lowered on a line, several hundred feet below the surface. Any sounds picked up by the transducer are sent back to the aircraft. This is done through a transmitter located in the buoy.

FIXED SONAR is another method for using active sonar, Fig. 22-23. In this way, the transducer is attached to the vessel, usually a submarine. The sound waves are sent out from the ship and the echoes are received in the same location.

The oceans are largely unexplored areas. As studies of the oceans increase, sonar will be useful in determining underwater features.

SUMMARY

Acoustical communication involves sending sound. These sounds travel in waves. The waves are created by a disturbance in a medium with mass and elasticity. The distance a message travels depends on pitch and loudness. However, several acoustical communication devices will aid messages in traveling farther.

Many types of acoustical communication systems exist in our world today. One of the most popular is the stereo system.

A stereo system consists of a receiver, amplifier, playback unit, and speakers. Each component has a specific function to do. Records and tapes are also part of the stereo communication system.

Radar and sonar are two more acoustical communication systems. Radar is radio-based and used to detect distant objects. Sonar works in a manner similar to radar. However, sonar detects underwater objects.

KEY WORDS

All the following words have been used in this chapter. Do you know their meanings?

Active sonar, Cartridge, Cartridge recorder, Cassette recorder, Compact disc, Continuous wave radar, Detection, Diaphragm, Doppler Effect, Dynamic speaker, Elasticity, Fixed sonar, Mass, Microphone, Microwave landing system, Momentum, Open-reel recorder, Passive sonar, Playback head, Pulse radar, Radar, Records, Sonar, Sonubuoy, Stationary heads, Stereophonic recording, Storing, Stylus, Tapes, Tonearm, Transducer, Turntable, Tweeter, and Woofer.

TEST YOUR KNOWLEDGE—Chapter 22

(Please do not write in the text. Place your answers on a separate sheet.)
MATCHING QUESTIONS: Match the definition in the left-hand column with the correct term in the right-hand column.

1. __ The ability of a body to return to its original shape.
2. __ Sounds that cannot be heard by humans.
3. __ The amount of matter that forms an object.
4. __ Building of strength and speed.
5. __ Frequency of a sound.
6. __ Strength of a sound.

   a. Pitch.
   b. Mass.
   c. Momentum.
   d. Elasticity.
   e. Inaudible.
   f. Loudness.

7. List three common uses of acoustical communication.
8. What invention led directly to the invention of the telephone?
9. Low-frequency sounds are sent out through the _____. High-frequency sounds are sent out through the _____.
10. Explain the types of radar.
11. What is the Doppler Effect? Give an example.

ACTIVITIES

1. Ask your school nurse to conduct a hearing test for your class. How does your hearing test? Also, ask the nurse to show the class how the testing machine works.
2. Bring in articles from magazines and newspapers on trends in acoustical communication.
3. Collect a variety of records and tapes, both old and new. Compare them to one another. Can you see any improvements from the oldest record to the newest record? What are some of the most obvious improvements?
4. Ask a stereo store sales rep to explain the latest in tuners and speakers.

<table>
<tr><td>

23

</td><td>

Broadcasting

</td></tr>
</table>

The information given in this chapter will enable you to:
○ *Explain the daily operations of a typical broadcasting station.*
○ *Identify the types of equipment used to broadcast TV and radio programs.*
○ *Describe the purpose and duties of various personnel at a broadcasting station.*

Broadcasting has become a major part of everyday life for people around the world, Fig. 23-1. Yet, it is a very young industry. Radio has only been around since the 1920s. Network television has only existed since the 1950s. And cable stations only became popular in the 1970s.

TV and radio programming provides hours of information and enjoyment each day. We stay informed of local and world events by watching news broadcasts. Weather warnings allow us to plan for bad weather. Sports programs provide entertainment. Game shows amuse us and challenge our knowledge. We hear current "hits" by listening to the radio, Fig. 23-2.

Commercial broadcasting often seems fairly simple to the common viewer or listener. However, the technical side of modern broadcasting is very complex. This chapter describes many of the techniques used to create TV and radio programs.

Fig. 23-1. Situation comedies (or "sitcoms") are a very popular type of television show. (Viacom International, Inc.)

INTRODUCTION TO BROADCASTING

Earlier we learned broadcasting includes the transmission of television and radio signals. These broadcasts take many forms. Some of the basic types of broadcasts include commercial, private, recreational, and federal/military.

Network stations (ABC, CBS, NBC) send out COMMERCIAL BROADCASTS. PRIVATE and RECREATIONAL BROADCASTS are sent and received by certain interested groups. CBs, ham radios, and walkie-talkies are used for private and recreational purposes. FEDERAL/MILITARY broadcasts are conducted by our government or the Armed Forces.

The remainder of this chapter will address commercial broadcasts. This is perhaps the most common form of broadcasting.

BASIC EQUIPMENT AND FACILITIES

Commercial TV and radio broadcasts use much of the same equipment. Also, they both are created in a STUDIO. This is the center of activity and attention. Most broadcast programming is recorded in the enclosed studio area. A radio studio is often smaller than a television studio, Fig. 23-3.

Both radio and TV studios are designed for the electronic equipment they must contain. Sound-proof glass and padded walls prevent noise from entering the sound studio. Support equipment is kept out of the way of announcers and newscasters. In TV studios, an attractive "set" hides ugly walls and fixtures, Fig. 23-4.

Next to most studios is the control room, Fig. 23-5. From this area, the producer can watch over programming as it happens. From here, shows can be edited, special effects added, or commercials inserted into the broadcast. The radio or TV signals leaving the control center go directly to the transmitter facilities.

Many types of electronic equipment are used inside the TV and radio studios. Perhaps the most common is a MICROPHONE (or "mike").

Fig. 23-2. This disc jockey is choosing some of the music he will play during his show. (Ball State University)

Fig. 23-3. Two types of studios. Left. A television studio must be large enough for a set, cameras, and workers. Right. A radio studio often has the controls, turntables, and microphone in one room. (Ball State University)

Microphones come in many forms. DIRECTIONAL MICROPHONES are placed in various positions around the studio. When only one voice is to be heard, a UNIDIRECTIONAL MIKE is used, Fig. 23-6. Announcers often use this equipment to quiet background noise. BOOM MICROPHONES can be suspended (hung) above the speaker. This is done to keep the microphone out of the way of the workers, Fig. 23-7. Other microphones include SHOTGUN and CARDIOID models.

Some types of equipment are used only in TV or only in radio broadcasting. For example, cameras, lights, and videotape recorders are found only in television stations. Devices found most often in radio studios include audio recorders and reel-to-reel tape equipment. This equipment will be discussed later in the chapter.

BROADCASTING PERSONNEL

PERSONNEL are workers at a business. Broadcasting firms employ many people. The most recognized job is that of the "on-air" personality, Fig. 23-8. These people work as actors, actresses, newscasters, and disc jockeys.

While their positions may seem glamorous, the workload is heavy. These people work even when they are not on the air. For example, television

Fig. 23-4. Walls, desks, posters, and carpeting in this studio hide bare walls and floors. (Ball State University)

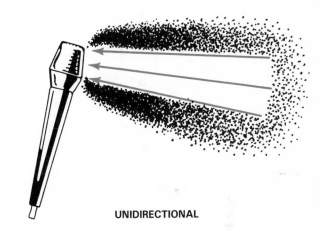

UNIDIRECTIONAL

Fig. 23-6. Unidirectional microphones pick up sound coming from one area in a room.

Fig. 23-5. Control rooms are used to monitor television and radio programs. (Ball State University)

Fig. 23-7. In the radio studio, boom microphones keep the control area clear. (Ball State University)

newscasters often research their own stories. At a small radio station, a disc jockey might also sell advertising time to local businesses.

Many on-air personalities have advanced degrees. Training in speech and journalism is common for all on-air staff.

In broadcasting, many people work "behind-the-scenes." The jobs done by these people are as important as the work of the on-air personality. The chief engineer is responsible for the equipment at the station. He or she must make sure all equipment works correctly. Without a chief engineer, the station would not be able to broadcast.

Radio talk shows require a producer to keep the show running. If calls are taken from listeners, the producer picks those calls that will get on the air. He or she also alerts the host when commercial breaks are coming. Talk shows often have additional personnel giving news and sports updates.

Other personnel prepare the script (or copy) for a broadcast. These people are known as SCRIPT WRITERS. The script is the text that will be read over the air. Some people write the news stories for announcers. Others create advertising slogans and feature items. These employees usually have some post-high school training.

Other behind-the-scenes personnel include sales staff and camera operators, Fig. 23-9.

COMMERCIAL RADIO BROADCASTING

Being organized is the key to success in broadcasting. This means all employees are working as a unit, toward a common goal. The common goal

Fig. 23-9. In order to do this job, post-high school training is needed. (Andy Johnston)

Fig. 23-8. The voices of these radio disc jockeys will be broadcast to their listening audience. (Ball State University)

is often to have the best (most listened to) station. To reach this goal, the station must be organized. Some people must be in charge. Others must be willing to listen. Most commercial radio stations are organized in the same way: ownership, management, and staff, Fig. 23-10.

STATION ORGANIZATION

Ownership

Many people own things. Some people own cars. Some own houses. Radio stations are also owned by someone. Often it is not one person, but a group of people. This form of ownership is called a CORPORATION. And many stations are owned by other companies. The owner holds the broadcasting license to the station.

Owners have the final word concerning the radio station. They set policies for the station. For example, the owner decides what type of advertising and programming will be used. They also make key decisions concerning money. This may include buying a transmitter or increasing the salary of a disc jockey.

Management

Radio station owners normally are not present at the station itself. They are not involved in the day-to-day work. Instead, they rely on the management group to run the station. A STATION MANAGER (or GENERAL MANAGER) heads the team. Other managers (or supervisors) direct each area of the station, Fig. 23-11.

The most important job of managers is deciding which market they want as listeners. A MARKET is a certain group of people that the station wants to attract. For example, adult markets are usually attracted to stations having more talk and less popular music by current groups. A teen-age market usually likes music stations that play "top 40" music. Foreign language stations specialize in delivering news to certain ethnic groups.

Once managers decide which market they want to attract, many more decisions are made. The biggest is whether to go all music, all talk, all news, or a combination. This is called the FORMAT. Format is the type of programs the stations will provide. Managers then decide what type of on-air personalities to hire. They also plan the broadcast day. The daily schedule is often shown as a PROGRAM CLOCK (or WHEEL), Fig. 23-12.

Staff

The staff actually performs the day-to-day tasks of the station. They include disc jockeys, newscasters, engineers, and receptionists, to name a few. The combined efforts of these people make the station successful. Depending upon the size of the station, staff jobs vary. Larger stations have different people for almost every task. Staff at smaller stations often perform more than one job.

PROGRAMMING AND SCRIPT WRITING

Programming involves deciding what kind of features will be broadcast. It also includes the time

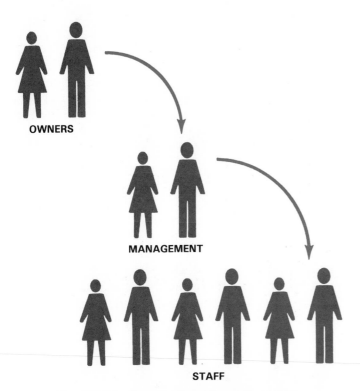

OWNERS

MANAGEMENT

STAFF

Fig. 23-10. Organization of a radio station.

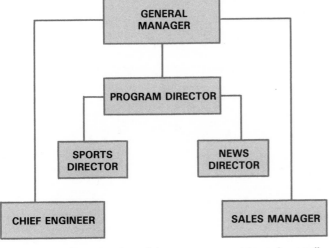

GENERAL MANAGER

PROGRAM DIRECTOR

SPORTS DIRECTOR

NEWS DIRECTOR

CHIEF ENGINEER

SALES MANAGER

Fig. 23-11. Organization of the management team in a radio station.

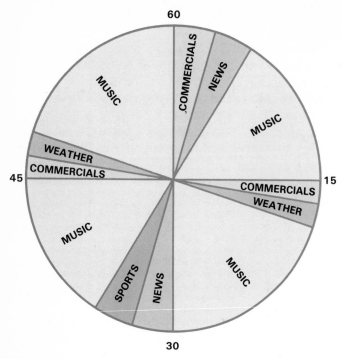

Fig. 23-12. This program clock shows an example of one hour of programming. What format do you think is used at this station?

different items featured throughout the day. Interviews and talk shows might be aired late in the day. Sports programs are broadcast live to most listeners. This may interrupt the "normal" schedule of the station.

SCRIPT WRITING is not used as much as it was in the 1930s and 1940s. At that time, before television was so common, radio stations broadcast popular shows. These were similar to television shows, without the video. However, there are still some radio shows being broadcast. National Public Radio (NPR) airs several programs of this type nationally.

Script writing is now used most often for commercials and public service announcements (PSAs). Many businesses ask the station on which they advertise to write their commercials. The business supplies information for the commercial. The CONTINUITY (kon-teh-new-ih-tee) WRITER works this into a script. Music and/or sound effects are often included. This same process occurs for PSAs, Fig. 23-13.

PRODUCTION TECHNIQUES

Production means getting all the needed items together in order to put a show on the air. Different formats need different production techniques.

A music show is produced with music as the top concern. Good sound quality is important. Therefore, the best playback equipment must be

they will be broadcast. Refer back to Fig. 23-12.

Programming is an important part of any successful radio station. This is because good programming attracts listeners.

The programming format varies with the station. Music stations can play rock-and-roll, jazz, or classical music. An all news format would have

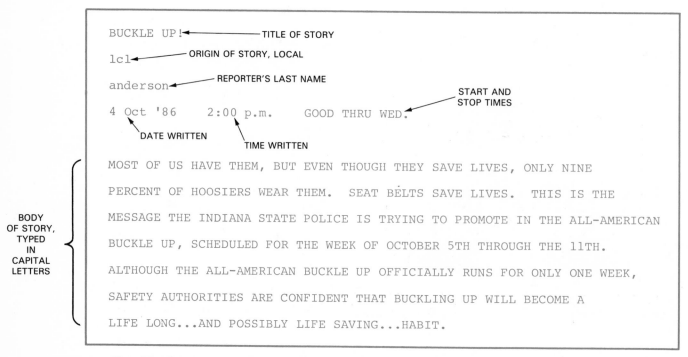

Fig. 23-13. An actual script used by radio announcers. (WBST, Muncie, Indiana)

used. Tapes or CDs (compact discs) are often used instead of records. This is because they hold up better under constant use.

Music shows also require a large music library. Usually a music show can be run by the disc jockey. DJs can play commercials, read the news, and play music without the help of others.

Standard radio work involves mixing audio sounds for transmission. Speech, music, and other sounds are picked up by microphones. These different signals are mixed in a console. All audible sounds should be recorded as close to the same level as possible. On the control panel, VU (volume unit) show the signal strength.

From the production area, the radio signal goes to the transmitter. The transmitter boosts the signal to the assigned frequency of the station. Radio transmitting antennas are often hundreds of feet tall. They are common sights along roadways. Transmission antennas are the devices which beam the signals to our home and car radios.

A great deal of work goes into a commercial radio broadcast. We have learned the basic work that is done. This work is exciting and important. Many people rely on radio broadcasts to stay informed. Radio broadcasting is an important part of the communications family.

COMMERCIAL TELEVISION BROADCASTING

Over 97 percent of homes in the country have television sets. This means television has some impact on most everyone's life. We have learned how television works and how the signals are transmitted. But how does a commercial TV station operate? We will learn about television broadcasting in the following sections.

Fig. 23-14. General managers must make many decisions.

STATION ORGANIZATION

Like commercial radio, television stations have three levels of personnel: owners, managers, and staff. However, the purpose and work in these areas are sometimes different.

Ownership

Many television stations are owned by the networks, ABC, CBS, and NBC. These large corporations are responsible for their local stations. Local stations are called AFFILIATES.

Many commercial stations are owned by smaller companies. These stations are not owned by a major network. They are INDEPENDENT STATIONS. Owners of independent stations often become more involved in their stations than network owners. This is because independent owners often own only one station.

Management

As in radio, the station manager, or general manager, is the top official in a TV station, Fig. 23-14. He or she is concerned with the entire station. The general manager sees that operating policies and FCC rules are followed. He or she also make major programming decisions. Often, these decisions are based upon years of broadcasting experience.

Working beneath the general manager are the managers of all the departments at the station, Fig. 23-15. These include sales, production, news, administration, and engineering.

The unusual part of television is that each show has its own organizational structure. A producer is normally in charge of the production of a certain program. It is the producer who brings the entire television show together. He or she makes up the budgets, supervises the project, markets the program, and manages the talent.

Staff

Personnel can be found working in five different areas of the station: administration, sales, news, programming, and engineering, Fig. 23-16.

Administrative staff includes typists, receptionists, and personnel directors, to name a few. They are responsible for the "front office" of the station. This group also includes the financial area. Accountants and bookkeepers keep track of the company money.

The sales staff is part of the marketing effort. These people sell commercial time to businesses. Other salespeople work with distributors of shows. Money raised by the sales staff is the major source of income for the station.

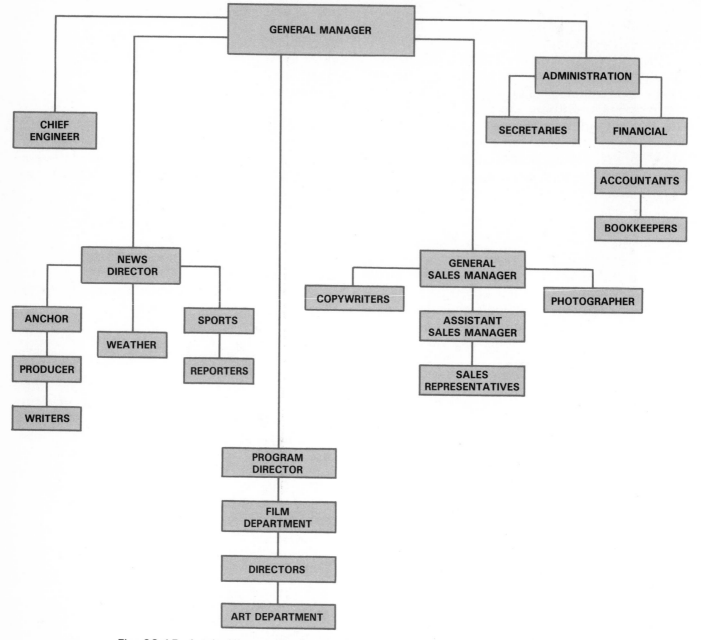

Fig. 23-15. A television station in a small market might be organized as shown.

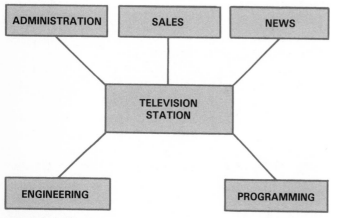

Fig. 23-16. Five areas of staff working in a television station.

All television stations feature news programs and specials. The station staff must collect and create stories for broadcast. A news department handles this task. Reporters and camera crews attend local events to record important news stories, Fig. 23-17. Back at the station, film and audio tracks are combined. The result is a "location" broadcast often seen on news programs.

The programming department produces the programs shown on the station. Writers and set-designers plan the show. Make-up artists and wardrobe assistants help with costumes and other details. Camera and sound personnel record the action on-stage. Then post-production staff finish the program for broadcast. This group includes

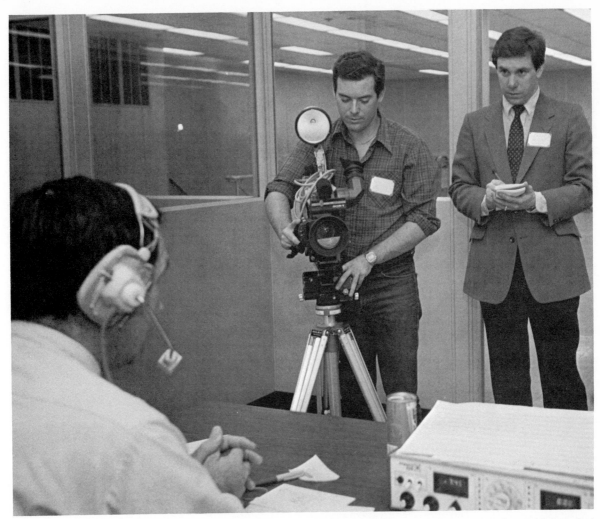

Fig. 23-17. This camera operator and reporter are covering a local news event. (Ball State University)

film editors and individuals who add "voice overs" to the program.

Engineers get the signals transmitted to viewers. There are a number of positions in this area. Maintenance engineers set-up, maintain, and repair equipment, Fig. 23-18. Remote broadcasts are handled by field engineers. Electrical engineers test and install new equipment at the station.

PROGRAMMING AND SCRIPT WRITING

In television work, programming involves planning and producing shows. These shows can come from three sources. They can be produced locally. They can come from the network. Or they can be obtained from an independent production group.

When shows are obtained from the network or an independent group, they are complete. The station needs only to "air" the show. They must be fit into the programming schedule.

Locally produced shows are done with the local audience in mind. Programming may include news

Fig. 23-18. Maintenance engineers test equipment to make certain it is working correctly. (Zenith Electronics Corp.)

features, children's shows, and local sporting events.

For programs produced in the studio, planning is done on a STORYBOARD. A storyboard serves as an outline for any audio-visual presentation. It is a series of rough sketches. Each sketch shows a certain camera shot. Video and audio directions are also given underneath each sketch. A storyboard form is shown in Fig. 23-19.

PRODUCTION TECHNIQUES

As we have already learned, production involves obtaining all items needed for putting a show on the air. The production techniques used in radio are not the same as those used in television.

Most television programs are created in studios. In these studios are lights, cameras, microphones, and sets, among other things. During production, personnel trained in various phases of production work the cameras, lights, and microphones.

In the following sections, we will discuss how these different pieces of equipment are used in television production.

Lighting and microphones

Lights and microphones are important production equipment. Lights create the amount of brightness needed to pick up an image. Microphones are used to catch all necessary sounds.

Both lights and microphones are kept out of the way of personnel during production. Lights are often hung from permanent (mounted in one place) supports. Microphones are put on the end of booms. With booms, they can be kept out of the camera range and used to follow performers as they move.

Two types of lighting are used in television production. One type is the basic light needed by cameras to pick up a picture. SPOT LIGHTING

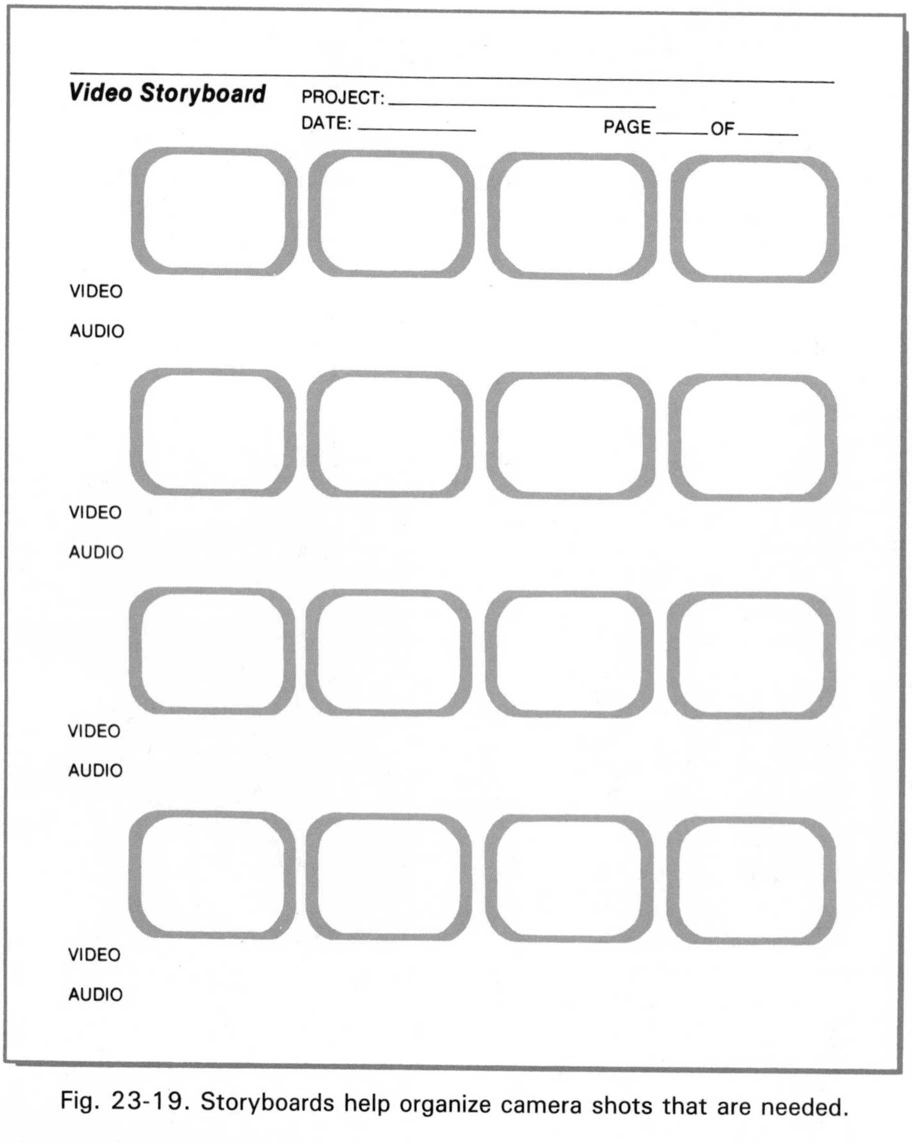
Video Storyboard PROJECT: ______

DATE: ______ PAGE ____ OF ____

VIDEO

AUDIO

VIDEO

AUDIO

VIDEO

AUDIO

VIDEO

AUDIO

Fig. 23-19. Storyboards help organize camera shots that are needed.

is used to brighten the set.

The other type of lighting used is CREATIVE LIGHTING. It is used to create a certain effect. For example, beaming a number of colors onto the stage provides an interesting visual effect.

Cameras

Perhaps the most important piece of equipment in the television studio is the camera. Cameras transmit the video information in television broadcasts. Large floor units are used in major television studios. Hand-held models are useful for remote broadcasts, Fig. 23-20.

Most productions require several cameras to record different angles. This allows the audience a number of views of any one scene. For example, on the evening news, you may notice the anchorperson turn his or her head to look into another camera. Switching camera angles is done to keep the audience interested. Imagine how bored you would become if only one camera angle was used for a full thirty minutes!

Like lighting, cameras are used in several ways. One type of camera viewpoint is used a great deal in newscasts. It is called REPORTORIAL. The newscaster speaks directly to the camera.

In OBJECTIVE viewpoint, the camera watches the action. This is the viewpoint used most often in television. Sports broadcasting is a common example of an objective viewpoint.

A third camera viewpoint is called SUBJECTIVE. In this instance, the camera takes the view of the actor or actress. Subjective viewpoint is used well in many dramas. Perhaps you recall seeing this used in a made-for-TV movie.

Special effects

Special effects are used to show an unusual event. Often these are natural occurrences, such as fire, rain, or fog. They are used to add to the setting. While some effects require costly equipment, many are simple to produce. For instance, dry ice or smoke is used to depict fog.

Audio special effects are easy to produce. Adding crowd noises or laugh tracks is a common practice. Strange music or tones help create excitement in horror films. Any sound can be added to a production during post-production work.

SET DESIGN

A SET is the area used as the location of the show. It includes all the equipment, scenery, and props needed for airing a show.

SET DESIGN is the organization of all these items. Cameras are placed where they may best record the action. Lights are placed in the proper location. Backdrops are constructed behind the stage. Set design creates an area for the production of the show.

Fig. 23-20. This lightweight video camera can be used on location broadcasts.
(Zenith Electronics Corp.)

Staging

If you were asked to build an area to look like your classroom, how would you do it? What kind of furniture would you pick? What kind of lighting would you use? Would it be bright or dim? Would you use plants? Posters? Bulletin boards?

Putting all of these items together is known as STAGING. Staging sets a feeling of a certain

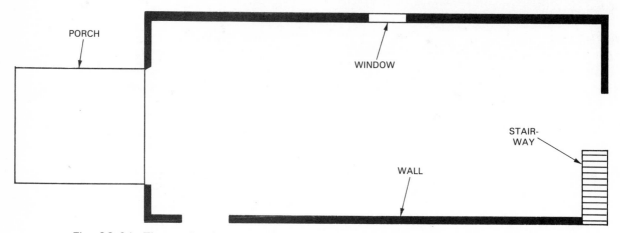

Fig. 23-21. The setting for a kitchen includes the walls, windows, and a stairway.

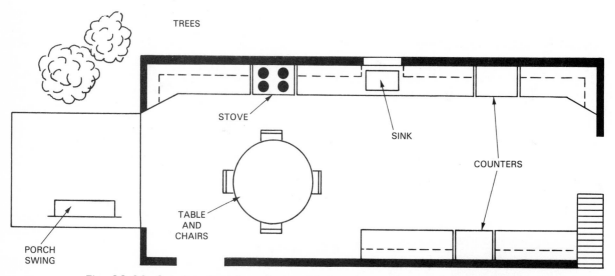

Fig. 23-22. Set dressings have been added to the setting used in Fig. 23-21.

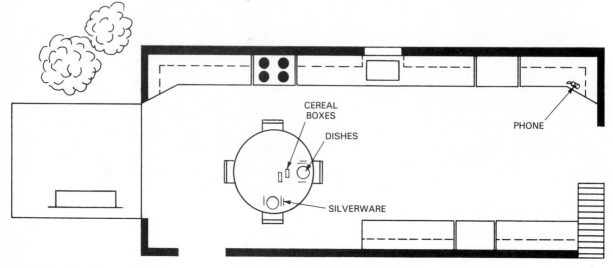

Fig. 23-23. Cereal boxes, silverware, and dishes have been added as props.

238

place. The idea is to make it look as authentic (real) as possible.

Scenery helps a great deal to make an area seem real. There are three types of scenery: settings, set dressings, and props.

SETTINGS are the large pieces that make up the areas of the set. For example, a setting for one show might include a kitchen wall, a porch, and a stairway, Fig. 23-21. Settings are made with a particular show in mind.

Furniture is the major part of SET DRESSINGS. However, also included are other items such as trees, books, lamps, and carpets. Set dressings include items that might be found in the location being shown, Fig. 23-22.

PROPS are those items actually used during the acting portion of the show. Telephones, tissues, purses, and food cartons are all props, Fig. 23-23. Props are different from set dressings because they are used by the actors and actresses while they perform on the set.

SUMMARY

In this chapter, we have covered the basics of modern broadcasting. This is an exciting industry. Radio and television entertainment is a part of our daily lives.

The production of a program is a complex task. Many resources are needed: people, equipment, and creative material (scripts). Radio and TV programs come from studios. Announcers or actors and actresses create the program for broadcast. Microphones and cameras "capture" the action on tape or film. The programs are then prepared for transmission by the station personnel.

It takes many talented individuals to operate a broadcasting station. Managers bring together the work of many staff members. Most individuals work "behind the scenes" developing programs. They support the familiar "on air" personalities.

KEY WORDS

All the following words have been used in this chapter. Do you know their meanings?

Affiliates, Boom microphone, Continuity writers, Corporation, Creative lighting, Directional microphones, Format, General manager, Independent stations, Market, Microphone, Objective viewpoint, Personnel, Program clock, Programming, Props, Reportorial viewpoint, Script writers, Set, Set design, Set dressing, Setting, Special effects, Spot lighting, Staging, Station manager, Storyboard, Studio, Subjective viewpoint, and Unidirectional microphones.

TEST YOUR KNOWLEDGE—Chapter 23

(Please do not write in this text. Place your answers on a separate sheet.)

1. Both TV and radio broadcasts are created in the _____.

MATCHING QUESTIONS: Match the definition in the left-hand column with the correct term in the right-hand column.

2. __ Types of programs a station broadcasts.
3. __ Group of listeners a station wishes to attract.
4. __ People who create the script for a broadcast.
5. __ Person who directs an entire radio or TV station.
6. __ A local station owned by a major network.

    a. General manager.
    b. Script writer.
    c. Format.
    d. Affiliate.
    e. Market.

7. What is involved in programming?
8. Name three sources from which television shows can be obtained.
9. _____ are an outline for any audio-visual presentation.
10. Explain the differences between settings, set dressings, and props.

ACTIVITIES

1. Watch a TV comedy show. List the props used by individuals on the program.
2. Plan, write, and record a 10 minute news/weather/sports program for your school.
3. Develop a program clock for your favorite AM and FM radio stations.
4. Build a model of a broadcasting set for a news program, game show, and comedy program.
5. Take a portable video camera to a local sporting event. Record the action on videotape.
6. Visit a local broadcasting station to watch a show being taped and produced.
7. Storyboard a video presentation describing your favorite hobby or game.

24 | Computers and Data Processing

The information given in this chapter will enable you to:

- *Explain the uses and purposes of computers and data processing.*
- *Identify and explain the differences between hardware and software.*
- *List the parts of computer hardware and how they work.*
- *Use a variety of input and output devices.*
- *Identify the types of software.*
- *Write simple programs in BASIC.*

Computers are a part of daily life. They are used in grocery stores, in classrooms, and in hospitals. Computers also come in many forms. A hand-held calculator is a computer. The machines that keep satellites in their orbit are also computers. Many of you may come in contact with computers every day at school. Some may be found in the library for student use. Others may be used in the office to prepare reports or budgets, Fig. 24-1. In this chapter, we will discuss what a computer is and what makes it work.

COMPUTER BASICS

A computer is a machine that accepts, stores, alters, and transfers information. The information is often called data. DATA is actually information that is not organized. This data is input to a computer. It is processed into usable form. The organized data is then released as output data. This is how DATA PROCESSING got its name.

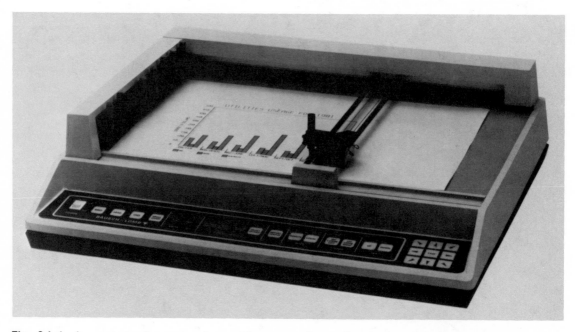

Fig. 24-1. A computer has many types of outputs. Depending on its use, computer output could be a pie chart, a letter or even a technical drawing. (Houston Instruments)

USES AND PURPOSES

Computers serve many purposes in processing data, Fig. 24-2. In an office, they can do word processing. WORD PROCESSORS are computerized, electronic typewriters, Fig. 24-3. They have a display screen and a limited memory (storage).

> COMPUTERS CAN DO:
> **WORD PROCESSING**
> **MATH CALCULATIONS**
> **SPREADSHEETS**

Fig. 24-2. Some typical office uses of the computer.

Fig. 24-3. This word processor can be used to type, print, and copy data. (A.B. Dick Company)

Word processors are often used to type simple letters that are sent to many people. These letters are called FORM LETTERS. The letter is typed once in total. As the letter is typed, it is displayed on a screen. Typing errors can be corrected. Entire paragraphs can even be switched around. Once the letter is completed, it is stored in the memory of the processor. Whenever a copy is needed, the name and address of the receiver are typed in the machine. The rest is done from computer memory.

Computers are also used to do math calculations. A business might calculate the total dollars of sales for a month. In an engineering company, complex math problems might be done on a computer. In a factory, boxes being sent out might be counted by a computer.

Computers are also used to process numerical data (numbers). This might take the form of a spreadsheet. A SPREADSHEET is a grid made up of rows (going across) and columns (going down) on a page, Fig. 24-4. It is used to keep information organized.

Many computer systems are designed to operate spreadsheet programs. Accountants use spreadsheets to keep the finances of their clients in order. Inventory control workers use spreadsheets to maintain a current count of products on hand.

We mentioned several ways that computers process words and numbers. These, however, are not the only uses. For example, computers are used to detect mechanical problems in a car, Fig. 24-5. After the mechanic examines the car, data is input to the computer. This data is processed. The output informs the mechanic where the problem is located.

Computing and data processing are exciting areas of technology. Through their use, people, businesses, and schools are able to better communicate.

BOOK NUMBER	TITLE	BOOKS IN WAREHOUSE	BOOKS AT PRINTER	TOTAL
0001-0	Basic Arithmetic	200	100	300
0001-1	Beginning Algebra	354	100	454
0001-2	Geometry	268	50	318
0002-0	English Composition	401	150	551
0002-1	Grammar and Spelling	183	250	433

Fig. 24-4. Spreadsheets are very helpful in organizing finances. This one shows the number of books in the inventory of a publishing company.

Fig. 24-5. Engine analyzers can be used to check for mechanical problems in a car.

HARDWARE AND SOFTWARE

The terms hardware and software are often used in data processing. HARDWARE is the equipment that makes up an entire system. It refers to the pieces of equipment: monitors, keyboards, and so forth. SOFTWARE is a set of instructions used to run the computer. These instructions are used for doing various processing. Software is also known as PROGRAMS.

Computers need both hardware and software in order to work. The hardware is combined with the software to process data. Think of hardware and software as a car and driver. The car (tires, seats, engine, etc.) is the solid piece of equipment. All the items are present for driving down the street. But a driver is needed to do the driving. The driver is like computer software. The driver directs and controls the car. In the same way, software helps the computer become a useful device.

MAINFRAME AND MICROCOMPUTERS

Two types of computers are often used today. They are the mainframe and the microcomputer.

A MAINFRAME COMPUTER is a large-scale machine, Fig. 24-6. It has a very large memory and is used mostly for general purpose work. General purpose work includes jobs such as controlling the payroll of an entire company. Because of the large memory, a mainframe is able to process all the data needed to control payroll. Mainframe computers have many uses in business and industry.

MICROCOMPUTERS, or PERSONAL COMPUTERS, are small, self-contained units, Fig. 24-7. Data can be input, processed, and output all on one system. This type of system often fits on a tabletop. Microcomputers are often used for per-

Fig. 24-6. Mainframe computers have powerful memories. This enables them to complete a variety of tasks. (Sperry Corporation)

sonal purposes. They have the same features as a mainframe, but on a smaller scale. To make up for less memory, however, microcomputers can be hooked up to larger mainframes. They then have access to the memory of the mainframe.

Computers have a countless number of uses in the Information Age. There are very few tasks that cannot be done using computers. Therefore, it is important to know the basic workings of computers. We will discuss these in the remainder of the chapter.

COMPUTER HARDWARE

As we have learned, computer hardware is the equipment that makes up an entire system. It includes both the computer and any attachments. These attachments are called PERIPHERALS. Examples of peripherals include printers, disk drives, and joysticks.

All systems have four basic pieces of equipment. They are the input/output devices, memory, storage, and the central processing unit (CPU).

CENTRAL PROCESSING UNIT

The CPU is the brain of the computer, Fig. 24-8. It is here that all data is processed. In addition, all basic computer functions are done in the CPU. One example of a function is turning the computer on and off. Another example is informing the user when the memory is full.

A CPU has two parts, the CONTROL UNIT (CU) and the ARITHMETIC LOGICAL UNIT (ALU). The CU directs the processing of each instruction. Signals are "steered" along the proper electronic pathway. The ALU actually performs the instructions. This may mean, for example, calculating numbers. A computer can only add and subtract numbers. Therefore, division is done by subtracting. Multiplying numbers is done by adding numbers several times.

A microcomputer's CPU can be contained on one small computer chip. A COMPUTER CHIP is a piece of silicon about 1/4 in. square, Fig. 24-9. The chip contains thousands of transistors. A CPU contained on a single chip is called a MICROPROCESSOR.

Fig. 24-8. Central processing unit of a microcomputer.

Fig. 24-7. Microcomputers are used in smaller businesses and by individuals. Are you familiar with these types of computers?

Fig. 24-9. A single computer chip is housed in this box. The small metal legs attach the unit to the computer.

MEMORY

The memory of a computer holds information for use by the CPU. Computers have two types of memory: random access and read-only.

RANDOM ACCESS MEMORY (RAM) contains information for use by the CPU. Software loaded into a computer goes to the RAM. This information is stored only while the computer is turned on. When the computer is shut off, the program is erased from the memory. Any input data is also erased, unless it has been "saved." Data is often saved on computer tapes or disks.

READ-ONLY MEMORY (ROM) is installed at the factory. It cannot be erased and is difficult to change. Instructions in ROM can only be "read" by the CPU. Programs or data cannot be stored in it.

ROM contains commands that allow a computer to function properly. For instance, when you press the letter "t" on a keyboard, a "t" appears on the screen. This happens because instructions in ROM send the proper signal to the monitor.

STORAGE

Often, data needs to be stored for a length of time. In these cases, the data is placed on a storage device. There are many kinds of storage devices now in use. Tapes and disks are used most often with personal computers, Fig. 24-10.

Microfiche and microfilm are used for long term storage. MICROFICHE is a copy of data reduced onto a four inch by six inch filmsheet. MICROFILM is a copy of data reduced onto 16mm continuous film.

INPUT/OUTPUT DEVICES

As we have learned, devices hooked onto a computer system are called peripherals. Many allow data to be put into the computer. Others are used to get information out of the system. Therefore, most peripherals are referred to as input/output devices, or I/O DEVICES.

An INPUT DEVICE is any piece of hardware used to input data into a computer. The most common example of an input device is a computer keyboard, Fig. 24-11. Other examples include card readers, joysticks, and light pens. Many computers get data over telephone line, through an input device called a MODEM. Disk and tape storage systems are also examples of input devices.

An OUTPUT DEVICE allows the user to get information out of a computer. This is done when data needs to be studied or recorded. A computer printer is a common output device. Other examples include video monitors, storage units, and audio speakers. Have you ever heard a computer "talk?" The output is an audible sound, rather than text on a screen or printer.

Input/output devices are necessary for working with computers. They allow human-to-computer communication. The computer would be useless without the option to input and output data.

COMPUTER SOFTWARE

Earlier we learned that computer software is a set of instructions used to run a computer. In this section, we will discuss the languages in which software is written and the types of software used. We will also explore basic programming techniques.

Fig. 24-10. Tapes and disks are the most common forms of storage used with computers.

Fig. 24-11. A computer keyboard. The card at the top of the keyboard informs the user of other functions of the keys.

LANGUAGES

In Chapter 20, we said that computers have to understand what is being said, in order for them to work properly. Computer languages allow this understanding.

A COMPUTER LANGUAGE is a set of commands that the computer can use to work with data. When these commands are put together as a related set, a program is made. When the computer reads the commands, it changes them into a language it can use. This process is similar to the work done by an interpreter.

An interpreter translates information between people who do not speak the same language, Fig. 24-12. It is important that the interpreter understands the languages being spoken. He or she can then act as the "go between," allowing these persons to communicate. Interpreters work in the same way as a CPU.

Like spoken languages, computer languages have a VOCABULARY (words) and grammar rules, Fig. 24-13. Some common computer languages include:

- BASIC (Beginner's All-purpose Symbolic Instruction Code). It is used in most personal computers.

- COBOL (Common Business Oriented Language). As the name states, COBOL is used a great deal in business.
- FORTRAN (FORmula TRANslator). FORTRAN is used widely for science and engineering programs.
- PL/1 (Programmer's Language 1). PL/1 is also used for science and business programs.

TYPES OF SOFTWARE

In a computer system there are usually three types of software. These three types are system, application, and utility, Fig. 24-14.

SYSTEM SOFTWARE consists of the operating system, language translators, and some utility programs. The OPERATING SYSTEM software is used to control the computer and its input/output and storage devices. It operates the hardware using a set of computer programs. These programs control all the other programs used on the computer. The operating system decides which program the CPU will work on at any time.

LANGUAGE TRANSLATORS are programs that change languages inside the computer. This is a needed function because the programs are often written in a language the CPU does not understand. So, the translator changes the information for the CPU.

UTILITY SOFTWARE, or service software, performs several jobs. HOUSEKEEPING is one of those jobs. Housekeeping consists of clearing storage areas, starting programs, and storing data for later use. It also instructs peripherals how to operate.

Another job of utility software is keeping track of computer activity. This activity is measured using PERFORMANCE STATISTICS. The measurement tells the user how well the computer is working.

A third job of utility software is for user convenience. For example, most computers have

Fig. 24-12. An interpreter acts as a "go between" for people who speak different languages.

```
NO, FAIL = NO, FAIL + 1
IF NO, FAIL    1000 THEN RETURN
IF NO, FAIL = NO, PAGE*55 THEN _
    NO,PAGE = NO,PAGE + 1 :_
    LPRINT CHR$(12)

NO,GOOD$(NO,FAIL) + CHANGED, ID$
```

Fig. 24-13. This portion of code is written in BASIC.

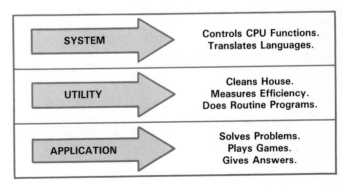

Fig. 24-14. Three types of software and their uses.

programs that will sort a file or merge (bring together) two files. The programs that allow users to edit work on the screen are also utility programs, Fig. 24-15. These "user friendly" utility programs do routine jobs. Without them, users would spend a great deal of time and energy doing it for themselves.

APPLICATION SOFTWARE does the data processing rather than control functions. These programs do the actual work, Fig. 24-16. They solve problems, play games, and give answers. Application programs are written to reach a certain outcome. Application software can be bought already programmed for personal use.

PROGRAMMING TECHNIQUES

To be a good programmer, good program design is needed. PROGRAM DESIGN is the path a program follows to reach the desired outcome. There is usually more than one way to design a program. However, one design is usually better than any of the others. A good programmer will decide which method is best and use it. This process is called the DESIGN PHASE. Programmers use various tools, or PROGRAMMING TECHNIQUES, to help them in designing software.

FLOW CHARTS

One common technique is the FLOW CHART, Fig. 24-17. It can be used to aid in coding (writing in a certain language) the program. Flow charts can be done at two levels: general and detailed.

A GENERAL FLOW CHART gives major functions and logic needed in a program. Often this is too simple to aid in coding. A DETAILED FLOW CHART, however, is used to give specific points of a program. This type of flow chart will match the actual logic that will be used in the completed program. Fig. 24-18 gives the American National Standards Institute (ANSI) meanings of the symbols used in Fig. 24-17.

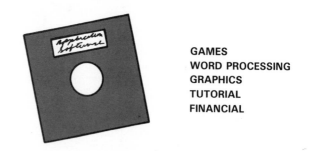

GAMES
WORD PROCESSING
GRAPHICS
TUTORIAL
FINANCIAL

Fig. 24-16. Some common application software available to consumers.

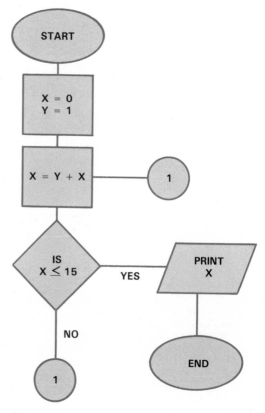

Fig. 24-15. Editing functions allow the user to replace, reorganize, or reword existing copy. If this function was not available, users would have to retype any changes.

Fig. 24-17. A general program flow chart.

BASIC INSTRUCTIONS

BASIC instructions are another type of programming technique. Like a utility program, they save time for the programmer. By using a certain word, an entire function is done.

BASIC instructions are either EXECUTABLE or NON-EXECUTABLE. Executable instructions are program instructions. They tell BASIC what to do while a program is running. Non-executable instructions cause no change in program flow or in the running of the program. Fig. 24-19 shows some examples of BASIC instructions.

RESERVED WORDS

Certain words have special meanings in certain languages. These words are known as RESERVED WORDS, Fig. 24-20. They include all commands, statements, function names, and operator names. Reserved words often send the program in a certain direction. If they were used in any other way, the program would most likely fail to work. This is because, except as reserved words, they have no meaning.

WRITING A COMPUTER PROGRAM

We have learned that computer programs can be written in any number of languages. But no matter what language a program is written in, there are certain steps to follow when writing. These steps provide the programmer with an outline to follow.

Often, the first step in writing a program is to write a general flow chart. This flow chart provides the basic requirements of the program. It is used to outline the logic.

After the flow chart is written, it is reviewed by both systems analysts and management. At this point, the general flow chart is broken down into modules (or sections). A detailed flow chart is developed for each module.

The detailed flow chart is used by applications programmers as an outline for the actual coding (instructions) of the program. A detailed flow chart has the same logical structure as the finished program.

The next step is to start coding the program. A program consists of a number of instructions that the computer understands. When these instructions are put in a certain sequence (order), the computer will do what you want. A program may consist of only one line or it may consist of 10,000 lines. Length is determined by the complexity (degree of difficulty) of the problem to be solved by the program.

The final portion of programming is typing in the code. Each line of code is given a number.

SYMBOL	SYMBOL NAME	MEANING
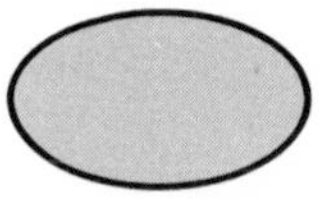 (oval)	TERMINAL	Start or end of an operation.
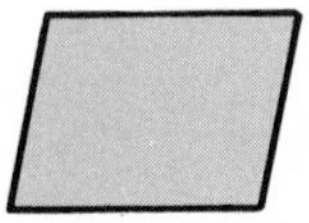 (parallelogram)	INPUT/OUTPUT	Input/output operation.
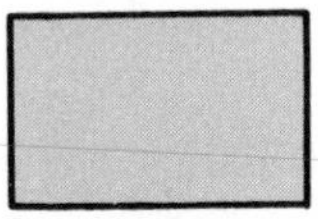 (rectangle)	PROCESS	A kind of processing function.
 (diamond)	DECISION	A logical decision to be made.
(circle)	CONNECTOR	Connection between parts of a flow chart.

Fig. 24-18. ANSI meanings of symbols used in flow charts.

DATA	Stores string or numeric data that are accessed by a READ instruction.
DELETE	Deletes program lines.
END	ENDS program execution.
FOR-NEXT	Executes a series of instruction in a loop, a given number of times.
GOTO	Branches to a specified line number.
IF	Tests a condition and changes program.

Fig. 24-19. Executable BASIC instructions.

LET	Assigns a value to a variable.
LIST	Lists the program that is currently stored in memory.
PRINT	Displays program data on the screen.
PRINT USING	Displays data according to a specified format.
READ	Reads a value from a data instruction and assigns to variable.
REM	Inserts comments into a program.
RENUM	Renumbers program lines.
RUN	Begins program execution.
SAVE	Saves a BASIC program on a diskette.
SQR	Returns the square root of a number.
TAB	TABS the cursor to a position on the screen.

Fig. 24-20. Examples of reserved words.

These line numbers tell the computer the sequence to follow when working the program. As when you count, the smallest numbers must come first.

It is important to know that the computer also knows how to count. So, even if you number lines out of sequence, the computer will process them in order, Fig. 24-21.

A common programming technique is to number lines in multiples of 10. In this way, if a line must be added at a later time, it is not necessary to renumber lines, Fig. 24-22.

"END" and "RUN" are two important commands in a program. "END" tells the computer that it has reached the end of the program. This command is usually followed by "RUN." "RUN" tells the computer to begin processing the program. This command is usually typed in separately from the code.

If a program is "clean," it will run correctly and you will receive the output you expect. Often, however, there are problems on the first few runs. These problems are called BUGS. The process of correcting these problems is known as DEBUGGING.

"SYNTAX ERROR" is a common bug. This means that, at some point in the program, language rules were not followed. Because the statement did not make sense to the computer, it could not be processed. This stops processing on the entire program.

A flow chart outlining the basic steps in writing a program is shown in Fig. 24-23.

SUMMARY

Computers serve many purposes. They can be used in the office, in school, or at home. In the

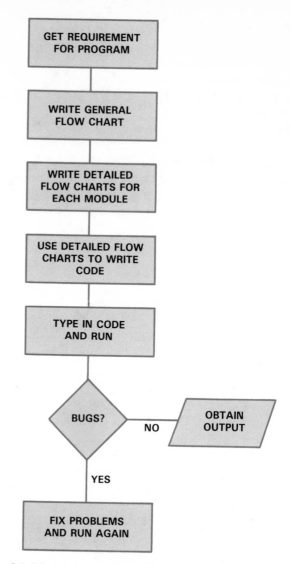

Fig. 24-23. Developing a program is outlined in this flow chart.

```
10 PRINT "COMPUTERS";
30 PRINT "FUN";
20 PRINT "ARE";
40 END
```

Fig. 24-21. The words in quotation marks are printed to construct a sentence.

```
10 PRINT "COMPUTERS";
20 PRINT "ARE";
25 PRINT "GREAT";
30 PRINT "FUN";
40 END
```

Fig. 24-22. This program is the same as in Fig. 24-21, except one line has been added. What happens in this program now?

office, computers can be used to type letters or control inventories. At school or home they can be used to type reports or to design a budget.

Two types of computers are commonly used in offices, schools, and homes. They are the mainframe and the microcomputer.

Mainframes are large machines with large memories. Microcomputers are desk-top size. They are used for smaller jobs: for personal use or for small business inventory.

Computers operate using hardware and software. Hardware is the actual system. It is composed of a central processing unit, memory, storage, and input/output devices.

Software is the programs that are run on the system. Programs are written in a language. Some common languages are BASIC, COBOL, FORTRAN, and PL/1. There are three types of software used with computers: system, application, and utility. Programs are written using techniques such as flow charts and reserved words.

When writing a program, certain steps are followed. These steps are used as an outline by management, systems analysts, and applications programmers.

Writing a general flow chart is the first step. From there, detailed flow charts are written for specific modules. Using the detailed flow charts, the program is coded and typed into the computer.

Once the program is completely typed, the computer is instructed to run the program. If the program is clean, the expected output is obtained. However, if the program has bugs, it will need additional work to run properly.

KEY WORDS

All the following words have been used in this chapter. Do you know their meanings?

Applications software, BASIC, BASIC instructions, Bugs, Central processing unit (CPU), COBOL, Computer, Computer chip, Debugging, Detailed flow chart, Flow chart, FORTRAN, General flow chart, Hardware, Housekeeping, Input devices, Language, Language translators, Mainframe, Microcomputer, Microfiche, Microfilm, Microprocessor, Output device, PL/1, Programs, Random access memory, Read-only memory, Reserved words, Software, Spreadsheet, Syntax error, System software, Utility software, Vocabulary, and Word processor.

TEST YOUR KNOWLEDGE—Chapter 24

(Please do not write in the text. Place your answers on a separate sheet.)
1. _____ _____ are computerized, electronic typewriters.
2. Explain the difference between hardware and software.

3. What do the following stand for?
   a. RAM.
   b. CPU.
   c. ROM.
   d. ALU.
4. Name three types of storage devices used with computers.

MATCHING QUESTIONS: Match the definitions in the left-hand column with the terms in the right-hand column.

5. __ Programs that do the data processing.
6. __ Words that have special meanings.
7. __ Changes information for the CPU.
8. __ Programming technique that shows the logic of a program.
9. __ Software used to operate the computer and input/output devices.
10. __ Software capable of housekeeping.
11. What are bugs?

a. Application software.
b. Language translators.
c. Utility software.
d. Operating system.
e. Flow chart.
f. Reserved words.

ACTIVITIES

1. Visit a computer store to learn more about personal computers. Collect literature on various models, made by a number of manufacturers. Learn about the different uses for the computer.
2. Visit your school's (or a nearby company's) data processing center, to see a working mainframe computer.
3. Design a class newsletter, using a microcomputer and a software package. Have it printed on the output device.
4. Develop a flow chart showing how to change a tire on a car. Be sure to use ANSI symbols.

Worldwide communication is provided by telephone
enterprises. (AT & T)

25 Communications Enterprises

The information given in this chapter will enable you to:
- ○ *Explain the process used when forming a communications enterprise.*
- ○ *Discuss the steps followed in establishing a corporation.*
- ○ *Diagram the structure of a corporation.*
- ○ *Explain the function of each department found within a corporation.*
- ○ *Discuss the closing of an enterprise.*

Are you familiar with the term enterprise? An ENTERPRISE is a business organization or company. The purpose of an enterprise is to make a profit from the sale of some product or service. If the money made from sales is greater than expenses, the company will be successful, Fig. 25-1.

There are many types of business enterprises. Examples include farms, shipping companies, and banks. In the communication industry, publishing and broadcasting businesses are quite common.

Fig. 25-1. The success of an enterprise is measured by the profit made. (Andy Johnston)

Many types of enterprises profit from the exchange of information in today's society.

This chapter describes the structure and management of communications enterprises. You will discover how companies are formed and financed. In addition, management of a business will be explained. By the end of this chapter, you should be able to form and operate a model communications enterprise.

FORMING A COMMUNICATIONS ENTERPRISE

Modern communications enterprises come in a large number of shapes and sizes. But, from the smallest printing firm to the largest computer company, most are formed in the same way.

First, the owners decide what to market (sell). Perhaps they have noticed a particular consumer (buyer) need or want. Maybe they have an idea for a product or service people may buy. In either case, the owners must decide how to produce the goods or service. Therefore, a goal is set for the company. All successful enterprises start with a goal.

The next step is to decide what form of ownership will be used, Fig. 25-2. The three most common types are proprietorship, partnership, and corporation.

TYPES OF OWNERSHIP

A business owned by just one person is called a PROPRIETORSHIP. The owner is known as a PROPRIETOR.

Proprietorships are formed when a business will be small and the capital (money needed to start and run the business) needs are little.

A PARTNERSHIP is formed when several people (two or more) own a business. A partnership has several advantages over a proprietorship.

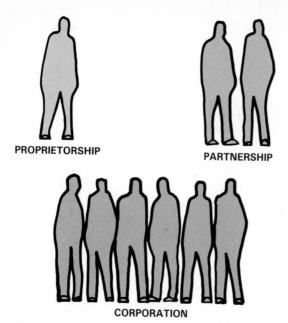

Fig. 25-2. There are three common forms of ownership.

Fig. 25-3. This bookstore is a corporation. That means many investors own a "share" of the business.

With several owners, a larger amount of capital is available. Also, more management skills are available.

The third form of ownership is the corporation. A CORPORATION is a legally formed business. This means a specific forming process is followed. (This process is discussed later in the chapter.) After completing this process, the business is seen as a corporation by the law.

Corporations are often owned by a large group of people, called STOCKHOLDERS. These people buy "shares" of the corporation. Their shares give them part ownership of the company. Many corporations have thousands of stockholders. Therefore, when one owner buys or sells shares, management is not usually affected.

Many communications businesses choose to form corporations because large amounts of capital are needed to reach company goals. Larger communications companies in the United States are corporations, Fig. 25-3.

OBTAINING RESOURCES

The ownership of the company has been formed. One of the first challenges of the new owners is to learn what resources are needed. The typical resources of an enterprise are money, people, materials, and knowledge, Fig. 25-4. These resources are needed for the company to work. In small companies, the owner (or owners) may be responsible for finding the necessary resources. However, in large businesses, the owners hire people to locate what is needed. For example, one

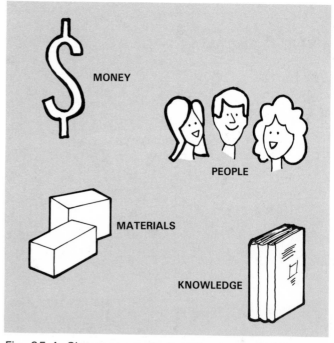

Fig. 25-4. Shown are typical resources needed in a successful communications enterprise.

activity includes hiring a management team to lead the company. They plan for and secure the company's resources.

The result of the wise use of resources is the production of goods or services. If we think of resources as INPUTS, then the final products are OUTPUTS, Fig. 25-5.

Can you think of items which may be considered outputs of communications enterprises? You are reading one! This book was produced by a type

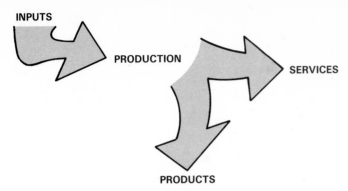

Fig. 25-5. The products or services resulting from communications enterprises are called outputs.

of communications firm: a publishing company. Other outputs include TV programs, computer software, billboards, school portraits, and telephone calls.

LICENSING A COMMUNICATIONS ENTERPRISE

In the United States, communications industries are highly regulated. This means that to form and operate an enterprise, there are several legal steps that must be followed. One of the first is that the owners must learn what agencies control their particular industry. For example, small printing firms may only require a vendor's license from the local government. However, telephone and broadcasting businesses are regulated by the Federal Communications Commission (FCC). They grant a commercial license to the company. Launching a telecommunications satellite involves gaining international approval from many governments.

Obtaining an FCC license is essential in the radio and television industry. This organization assigns broadcast frequencies and governs transmitter power and location. They also grant operating permits (licenses) to station personnel. These are granted through various testing procedures.

Once a station is "on the air," the FCC constantly reviews it. A broadcasting log (record) must be kept. This details what went on the air and when. In order to be granted a renewal of its operating license, a station must meet certain standards.

ESTABLISHING A CORPORATION

There are two main legal documents used when forming a corporation. They are the Articles of Incorporation and the corporate bylaws. Every corporation has to prepare and file these items with governmental authorities.

The ARTICLES OF INCORPORATION are an outline of the structure and purpose of the business. Among the important items listed on this charter include:

1. Name and address of the proposed company.
2. Purpose of the proposed corporation.
3. Names and addresses of the persons forming the corporation.
4. Location of the corporation office within the state.
5. Amount and kind of stock to be offered.
6. Names of principle officers of the company.

This information is supplied to the state government where the corporation is to be located. If approved, a charter will be granted to the enterprise. This allows the new organization to conduct business as a corporate body.

In addition, the BYLAWS of the enterprise must be drawn up. This document provides rules about the operation of the company. For example, it outlines the number and duties of company directors. It also describes the business practices of the firm. A list of those items that need stockholder approval is also included in the bylaws. Finally, important details related to the stockholders are covered. Voting procedures and stockholders rights are guaranteed by this document.

Upon completion of all legal requirements, the enterprise begins operation. A company organization (structure) is developed. The major groups of the business include personnel, finance, marketing, and production, Fig. 25-6.

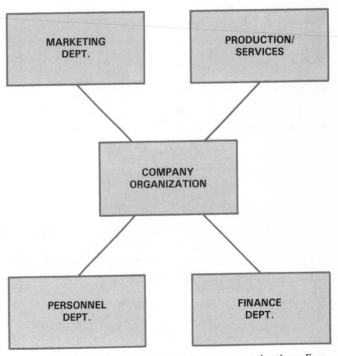

Fig. 25-6. The major departments in a communications firm.

STRUCTURE OF A CORPORATION

Many communications enterprises have a small staff that performs many tasks. Larger companies may have an entire department just to pay debtors and bill creditors. However, all enterprises usually have at least four activities in common. These include the finance, personnel, marketing, and production functions. In most corporations, each activity is done by a certain department. A management team directs their activities, Fig. 25-7.

CORPORATE MANAGEMENT

Most corporations' managerial structures contain several "levels of authority," Fig. 25-8. The owners have the final say in the running of the company. We know these people as the stockholders. However, all the stockholders cannot always be available to make key rulings. Therefore, their interests are covered by a board of directors.

The BOARD OF DIRECTORS make many of the policies for a corporation. They make their policies as they feel stockholders would. They also advise management on important issues.

The top person in the executive staff is called the PRESIDENT or CHIEF EXECUTIVE

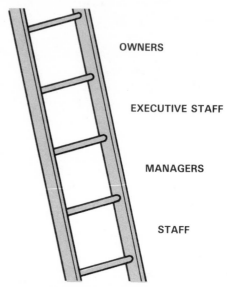

OWNERS

EXECUTIVE STAFF

MANAGERS

STAFF

Fig. 25-8. Common levels of authority.

Fig. 25-7. This manager is in charge of the computer operations center of a large corporation. (Santa Fe Railway)

OFFICER (CEO). This person directs the company's day-to-day work schedule. The president is responsible for all work of the business. Since this is a large task, a vice-president is usually in charge of each department. VICE-PRESIDENTS watch over the work done in their groups. Each vice-president reports to the president of the company.

MANAGERS have many functions in communications firms. First, they set goals for the enterprise. This includes planning and organizing the company's resources. Managers then direct the daily functions of the company. They also rate the work in progress.

The STAFF is made up of people hired to do specific jobs. For example, the staff at your school consists of secretaries, teachers, guidance counselors, and cafeteria workers. These people are hired because of their talents, to do a specific job. Corporations hire staff workers based on the needs of the company.

FINANCE

The finance department controls a corporation's legal and monetary affairs, Fig. 25-9. This includes raising capital (money) for the enterprise. Finance employees also keep track of money through record keeping activities, like budgeting and accounting. Paying taxes is another important task for this department. The finance group is also responsible for purchasing supplies. Supplies include items like typing paper, glue, pens, and stationery.

The finance department also often handles the legal matters of the company. This involves working with many government agencies. Obtaining an operator's license or charter is an important task. Maintaining legal paperwork is also vital to the company's affairs.

The financial department prepares many reports concerning the status of the company. These details are available to the public, in written reports. One kind of written report is the ANNUAL REPORT, Fig. 25-10. This report includes information such as stock price ranges, summaries of operation costs, and a note to the stockholders (owners).

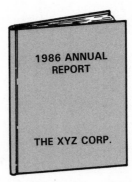

Fig. 25-10. Annual reports are prepared by the finance department and made available to all interested persons.

PERSONNEL

The personnel department has many tasks to do. Hiring and caring for the company's workers is one of the most important. The personnel department is also the company's link to the local community. Therefore, the personnel department is also in charge of public relations (PR). If a labor union represents workers in the company, the personnel department deals with the labor leaders.

The primary function of personnel is to bring together the workforce of the company. Personnel hires and maintains a good workforce. After all, people are the most important resource in any company, Fig. 25-11. All businesses need a variety of skilled workers and managers, Fig. 25-12. Each employee will add his or her own skills, knowledge, and experience to the business.

The steps taken when hiring employees follows a fairly common path, Fig. 25-13. However, several ways exist to recruit employees. For example, many firms make jobs known in local newspapers. You may see these "help wanted" ads in your hometown paper every day. Other enterprises announce auditions or tryouts. This is

Fig. 25-9. Functions of a finance department.

common in the entertainment industry. Many businesses let local or state employment agencies find employees for the firm.

Once ads have been answered, interviews completed, or auditions held, the best person for the position is hired.

Fig. 25-11. This employee performs various useful tasks for the company. (Ball State University)

Managers hire people they believe will give the most to the enterprise. For example, advanced training from a college or technical school may make one applicant more desirable than another applicant. Having a cheerful, pleasant personality is also desirable. And, of course, employers look for responsible, safe, and loyal persons when filling available positions.

PRODUCTION AND SERVICE

Communications enterprises are formed to sell an item or a service. This includes activities like publishing books, developing film, and transporting mail through computers. We will discuss separately those businesses that sell products and those that sell services.

TELEVISION STATION	
General Manager	Sports Reporters
Engineers	Camera Operators
News Director	Stagehands
Photographers	Sales Representatives
Producer	Audio Technician
Anchors/Announcers	Video Technician
Writers	Special Effects Director
News Reporters	Makeup Artist
Weather Reporters	Hairdresser

NEWSPAPER	
Publisher	Composing Staff
Editor-In-Chief	Librarians
General Manager	Classified Staff
Managing Editor	Press Operators
Advertising Manager	Photographers
Production Manager	Reporters
Circulation Manager	Copy Editor
Sports Editor	Secretarial Staff
News Editor	Sales Staff

RADIO STATION	
General Manager	Music Director
Sales Manager	Newscasters
News Reporter	Disc Jockeys (DJs)
Program Director	Audio Technician
Sports Director	Engineers
News Director	Sound Mixer
Script Writers	Sales Staff
Researchers	Music Librarian
Copy Editors	Secretarial Staff

Fig. 25-12. Positions in typical communications enterprises.

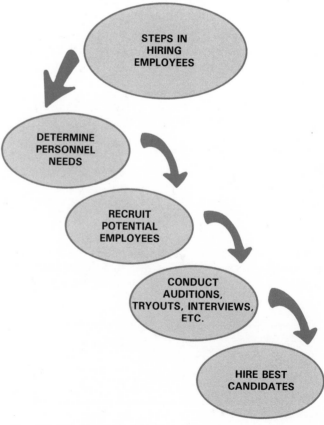

Fig. 25-13. Steps in hiring workers in an enterprise.

Businesses that sell products give buyers an item they can actually hold. For example, printing firms produce books, magazines, and newspapers. Larger companies assemble items like radios and television sets. Other communications firms produce and market albums, tapes, and computer disks.

Providing a service is quite different than providing an item. Telephone companies provide a service. Through the use of a telephone company's equipment, you are able to call friends. There is no physical product. They provide you with the materials necessary for making a phone call. This is a service. You pay for this service with your telephone bill. Many communications firms are service-related, Fig. 25-14.

The television industry is also service-related. Actors and actresses perform on various shows. Anchors on the local station report the day's important news. While you never hold any physical object, you are being entertained. This is a service. And like the telephone, you pay for this service. With cable TV, you pay a monthly bill. With "free" TV there are INDIRECT COSTS. For example, advertising time on a popular show is very costly. Despite the high cost, however, many companies will advertise during that show. In order to pay the advertising fee, companies often raise the price of their goods or services. This is an example of indirect cost.

MARKETING

Among other tasks, the marketing function involves getting consumers to buy a company's products or services. ADVERTISERS inform the public what new items are available. A SALES FORCE will complete the task of selling the goods or services to potential buyers. Then the DISTRIBUTION UNIT delivers the items to customers. This area may also include a marketing research division to access customer needs and desires. The major functions of a marketing area are listed in Fig. 25-15.

Marketing research is vital to any enterprise. For the company to succeed, it must understand consumer wants and needs. By surveying many

Fig. 25-15. Functions of a marketing department.

people, firms find out the desires of their audience. For example, researchers determine what TV programs are highly viewed. They also follow trends in entertainment and business. If they see, for example, that videos are very "big," they may style their ads in a video format. Marketing departments also measure the success of sales efforts.

All companies promote their products or services in some manner, Fig. 25-16. Common means of advertising include TV commercials, newspaper ads, and billboard displays. In Chapter 3, we learned that messages can inform, persuade, instruct, and entertain. Advertisements are messages. Therefore, they can serve any of those purposes. For example, a persuasive angle includes providing coupons or having sales.

Sellers and buyers in communications firms vary by industry, Fig. 25-17. For example, sales representatives (reps) for a cable company sell their TV service to local customers. Publishers sell books and magazines to individuals, stores, or schools. Large TV networks sell advertising time to clients. The marketing staff at a recording company sells records and tapes to stores. Others simply sell movie or concert tickets to consumers.

A marketing department is made up mostly of a sales force. These people contact potential buyers of the company's goods or services. A sales manager directs the company's sales force. The manager assigns areas (sales territory) to staff members. This area may include a geographic region or certain accounts. Experienced salespersons work with the larger customers.

SERVICE-RELATED COMMUNICATIONS INDUSTRIES	
CABLE SYSTEMS	TELEVISION
RADIO	VIDEOTEXT
TELEPHONE	SATELLITES

Fig. 25-14. All of these communications industries supply a service. Can you name any others?

Fig. 25-16. Advertisements appear in various forms, in different places. Do you recognize any of these common ads?

EXAMPLES OF PRODUCTS/SERVICES OFFERED TO CUSTOMERS	
COMPANY	PURCHASER BUYS
Stage Production Firm	Tickets to Concerts/Plays
Television or Radio Station	Advertising Time "On The Air"
Computer Software Firm	Complex Computer Programs
Publishing Firm	Books and Magazines Available to General Public
Film Production Business	Movies (Films) For Theater Showings
Recording Company	Popular Records and Tapes

Fig. 25-17. Communication products and services are purchased by a variety of customers.

Members of a sales force often have years of selling experience. Many spend large amounts of time traveling to clients and potential clients. Most salespersons are paid a salary for their work. In addition, they might also receive a share of each sale. This share is known as a SALES COMMISSION.

The marketing function also includes a distribution group. Distribution involves sending products to buyers. Again, this function varies by industry. Small companies often deliver their products or services directly to consumers. Large firms let other distributors sell their products. Businesses in the recording and entertainment industry have a complex distribution system.

MANAGING THE CORPORATION

Communication enterprises must be managed in an efficient manner. Managers direct the operation of the business. They serve as supervisors, directors, or administrators. Their knowledge and experience is vital to the success of the company.

Among the activities to be managed are the design, production, and transmission of messages. Hiring and training personnel is important. Also, financial affairs must be closely watched.

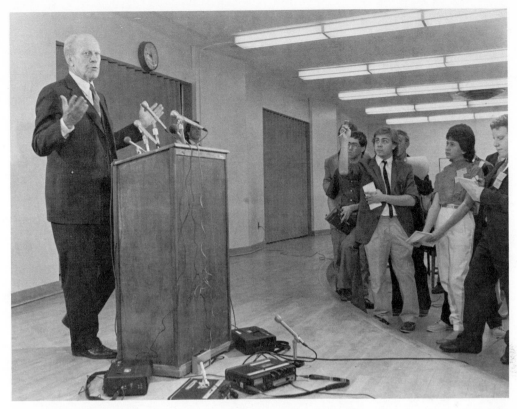

Fig. 25-18. These professional journalists are collecting information for their next radio newscast or newspaper article. (Ball State University)

DESIGNING PRODUCTS AND SERVICES

Products and services are designed by creative persons. Reporters write their news reports, Fig. 25-18. Artists and graphic designers lay out magazines, flyers, and posters. Stage crews produce the set designs for TV shows. Authors write the manuscripts for books. Photographers compose visual images for TV shows and motion pictures. Drafters develop complex drawings with the aid of computers, Fig. 25-19.

Design activities in communication industries are not limited to these few examples. Several other design activities are shown in Fig. 25-20.

PRODUCTION OF MESSAGES

Production tasks include recording, filming, and printing messages. Managers direct workers who do these jobs. For example, supervisors direct the operators of various production equipment, Fig. 25-21. The technical director at a radio station engineers a radio broadcast.

During production, managers check the progress of all work. They decide if changes are needed. This may include shifting resources or schedules. Also, managers decide if quality standards are being maintained.

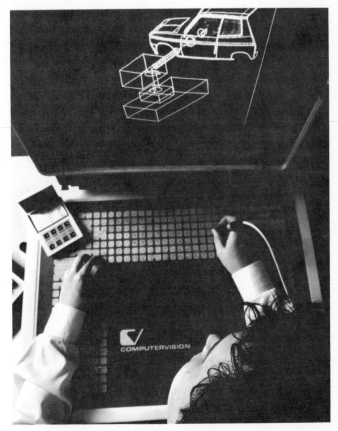

Fig. 25-19. Industrial designers often develop working drawings with the aid of computer systems. (Computervision Corp.)

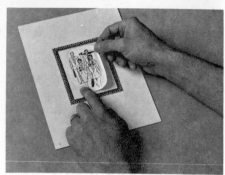

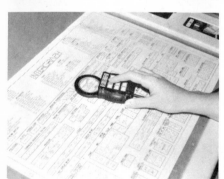

Fig. 25-20. Design activities in information industries.

TRANSMISSION OF MEDIA MESSAGES

Management is important in mass media. A great deal of messages are constantly sent. Messages are transmitted by audio, visual, and audio-visual means. There must be people in charge of organizing and reviewing this information.

Among the most familiar audio message is the radio program. Popular music and talk shows are transmitted to our homes or workplaces. Radio station managers decide what type of shows will be broadcast. The visual media include photographs and printed materials. Television and motion pictures are types of audio-visual communication, Fig. 25-22. Managers of theaters decide what films will be shown.

MAINTAINING FINANCIAL RECORDS

Financial records of businesses are often kept in computers, Fig. 25-23. These records include money earned and spent. The official record of all financial matters is a GENERAL LEDGER.

As we learned, company income is derived from the sale of products or services. This money is used to pay the expenses of the company, Fig. 25-24. For example, managers receive a salary for their efforts. Workers are paid for their productive time. Equipment and facilities are a major expense in most companies. Materials and supplies are often the most costly expenses. Utilities and other operating expenses must be paid, too.

After paying all the bills, the remaining money is called PROFIT, Fig. 25-25. A successful firm will return a profit to investors in the business. The profits are normally split among the owners of the firm. In a corporation, stockholders receive a DIVIDEND for each share of stock they own. The dividend is their share of company profits.

CLOSING THE ENTERPRISE

Many new corporations are formed each year. Others are terminated; they stop conducting

Fig. 25-21. This worker in a printing firm operates a variety of printing equipment. (Ball State University)

Fig. 25-23. Many companies use computers to keep the financial records of the business. This manager is reviewing the latest data of her company. (Sperry Corp.)

Fig. 25-22. In a theater, employees who work in the projection booth are known as projectionists. This projectionist is loading the feature film for the next showing. (Andy Johnston)

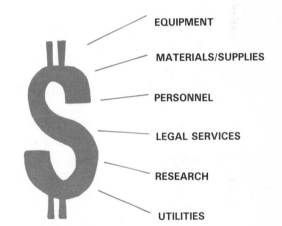

EQUIPMENT

MATERIALS/SUPPLIES

PERSONNEL

LEGAL SERVICES

RESEARCH

UTILITIES

Fig. 25-24. Typical expenses of a communications enterprise.

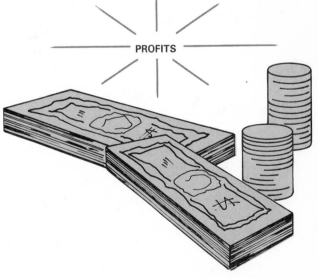

PROFITS

Fig. 25-25. The major goal of every enterprise is to make a profit.

business. Companies that fail to make a profit are often closed down by stockholders. Owners may then invest their money elsewhere. Due to financial or legal problems, the courts may order a firm to halt operation.

In either case, closing a corporation requires several legal steps. First, owners are informed of plans to close the company. Then a vote is taken at a meeting of stockholders. If the vote is approved, the corporation is dissolved.

Company officials then fill out the paperwork required to terminate the firm. The ARTICLES OF DISSOLUTION are filed with the federal government. Company assets are sold. ASSETS include equipment, buildings, and property. Extra materials are sold to other firms who use those materials. Employees are laid off from their jobs. Then all bills are paid in full.

Any remaining funds are distributed to the original owners. However, if the corporation has been in debt, there will probably not be any left-over money.

SUMMARY

Communications enterprises profit from the exchange of information or ideas. They produce materials or provide services to customers. Businesses in the communications area are known as information industries.

Information enterprises are often formed as corporations. These corporations include many owners, called stockholders.

Forming communications enterprises involves many legal steps. Approval is obtained from governmental agencies. A charter and/or license is granted to the firm. Then managers gather the needed resources to start the enterprise.

Corporations have many levels of authority. Managers direct employees who design, produce, and transmit messages. The company profits from selling products or services to customers.

KEY WORDS

All the following words have been used in this chapter. Do you know their meanings?

Advertiser, Annual report, Articles of Dissolution, Articles of Incorporation, Board of Directors, Bylaws, Chief Executive Officer, Corporation, Distribution unit, Enterprise, Indirect costs, Inputs, Levels of authority, Management, Marketing, Outputs, Partnership, Personnel, President, Proprietorship, Salary, Sales commission, Sales representative, Shares of stock, Staff, Stockholder, Termination, and Vice-president.

TEST YOUR KNOWLEDGE—Chapter 25

(Please do not write in the text. Place your answers on a separate sheet.)

1. What is the first step in forming a communications enterprise?
   a. Decide on the form of ownership.
   b. Determine what to market.
   c. File Articles of Incorporation.
   d. Choose a managerial structure.
2. A _____ is the only owner of a business.
3. Give two advantages of a partnership over a proprietorship.
4. Owners of a corporation are called _____.
5. Name four resources needed by most businesses.
6. Explain the process of establishing a corporation.
7. Most corporations have:
   a. Finance departments.
   b. Personnel departments.
   c. Marketing departments.
   d. Production departments.
   e. All of the above.
8. The term CEO stands for _____ _____ _____.
9. Job positions can be made known in three ways. Name these ways.
10. Explain how products and services differ.
11. What is the name of the legal document filed when closing an enterprise?

ACTIVITIES

1. Have your class start a communications corporation. Choose a name. Determine what you will sell. Take every step, from filing the proper papers and documents, to closing the enterprise.
2. Tour several local service- and product-related information industries. Compare their operations. List the products or services they offer to customers.
3. Interview the president of a communications enterprise. List the daily activities of this person.
4. Choose a major communications corporation. Follow the stock values for that company. Use companies listed on the New York Stock Exchange. Keep a written record of your company. How did it do over five days?
5. Collect the annual reports of major information firms. Have a class discussion about the various charts. What do they mean? How are these companies doing financially?

Communication— Today and Tomorrow

The information given in this chapter will enable you to:
○ *Identify current trends in communication.*
○ *Recognize possible developments in the future of communication technology.*

Technological growth has occurred rapidly in the last few years. Mathematicians describe this development as a GEOMETRIC PROGRESSION. Instead of a gradual (slow) increase in technology, growth has skyrocketed, Fig. 26-1. This steady growth pattern is called EXPONENTIAL GROWTH. It results from inventors building upon the knowledge and improvements of other inventors.

CURRENT TRENDS IN COMMUNICATION TECHNOLOGY

The area of communication has progressed rapidly. Traditional billboards have been replaced with electronic billboards, Fig. 26-2. These painted and lighted signs are quite lifelike. They are also very effective in attracting the attention of potential buyers. Colorful T-shirts can be made with ease due to the help of computer scanners that separate colors automatically. Automatic screen process presses allow the printing of four to eight color prints.

In lithography, new presses can print several colors at one time. A laser process can receive information through microwaves and produce a plate from the signals alone. This replaces negative filming and stripping procedures. Newer platemakers produce lithographic plates ready for loading onto the press immediately.

Photographic films are now available for unusual light or speed conditions. Very "fast" films (1600) are used to photograph sporting events or dark indoor events. "Slower" films are used for portraits or bright outdoor events. A range from 100 to 1600 ASA is available at most photography shops.

Cameras are becoming very easy to operate. Fully automatic cameras adjust the focus and set the f-stop electronically. Some cameras even "tell" the user if the light is sufficient to take a particular shot, Fig. 26-3.

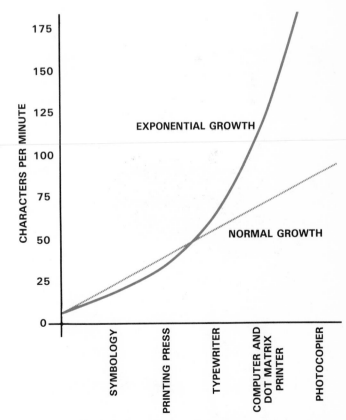

Fig. 26-1. Advances in communication technology happen at a rapid rate. The improvement in printing is shown as a solid line. The dashed line shows a normal rate of progress. Growth that occurs faster than normal is said to be exponential.

Fig. 26-2. Bright displays attract our attention as we pass by.

Photocopying machines now reduce and enlarge prints easily. They can also print on both sides, collate (sort), and pull data from computers. Laser printers are becoming popular in many industries.

"THE SCENE IS TOO DARK . . ."

Fig. 26-3. Modern cameras are a great aid to amateur photographers. The camera adjusts itself for different lighting conditions and distances.

Satellites are capable of keeping track of weather conditions. This is helpful when meteorologists (specialists in the study of the atmosphere and weather) need to follow the path of a storm. Satellites also are used to complete long distance telephone calls. They aid in ship navigation and surveillance, Fig. 26-4.

Many AM radio stations now broadcast in stereo. Until recently, only FM signals could achieve a stereo wave. Now car and portable radios will produce the same sound quality as a home stereo.

Home and office telephone service has changed a great deal over the past few years. Household telephones can now handle multi-party calls, call waiting, and multiple lines. Automatic dialing is another popular feature on many phones. Some business systems permit the transfer of pictures and financial data over normal telephone lines.

Videocassette recorders (VCRs) are found in more homes than ever before, Fig. 26-5. They come in a variety of models—front loading, top loading, VHS format, Beta format, cable ready, etc. Movie rental for home VCRs is now a multi-million dollar business.

Perhaps the greatest progress in communication

Fig. 26-4. Communication satellites allow us to send messages around the globe at the speed of light. They are the basis of worldwide telecommunication networks. (Ford Aerospace & Communications Corp.)

Fig. 26-5. Today's videocassette and disc players are current advances in electronic technology. (RCA)

technology has been with computers. Lower costs and improved technology have made computers available to nearly everyone. Individuals and businesses alike use computers to maintain day-to-day operations. Computer-aided design (CAD) is popular with designers and engineers. CAD is used to quickly and accurately design and draw

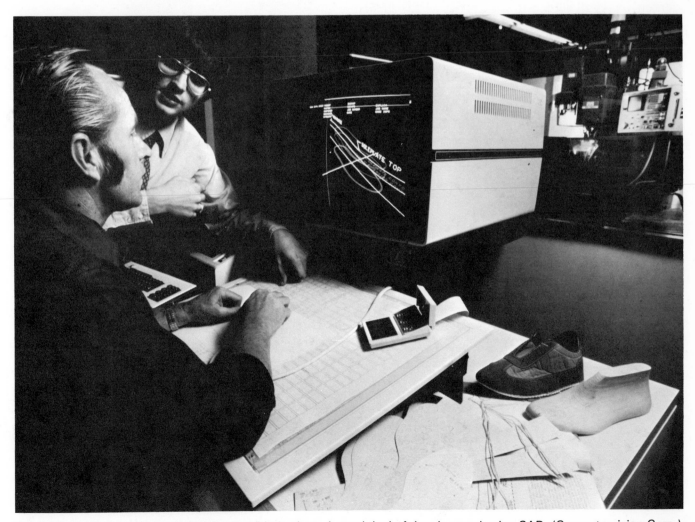

Fig. 26-6. CAD is used here to design an athletic shoe. A great deal of time is saved using CAD. (Computervision Corp.)

house plans, produce blueprints, etc., Fig. 26-6.

Computers appears in many forms. In cash registers, they total grocery bills. Placed in wristwatches, they keep time for a number of years. And in cars, they watch over the functioning of automobiles. Electronic sensors inform the driver how the car is operating, Fig. 26-7.

COMMUNICATION TECHNOLOGY OF THE FUTURE

Researchers make accurate forecasts about the future by reviewing past trends and current technologies. Many researchers believe that the future in communication technology is limited only by our imaginations. If an idea or product is important enough, technology can be developed to make it a reality.

TELEVISION

Televisions in the future will be much more than a box. They will be complete home entertainment centers, Fig. 26-8. Some televisions will have a split screen. A small screen in the corner of the larger screen will make it possible to watch two television programs at once. A zoom lens would make it possible to enlarge a portion of a picture on the screen. This would be handy for studying a golfer's swing or a tennis player's grip. Other innovative ideas are TVs with a "freeze frame" capability and TVs that double as audio-visual telephones. It would be possible to see as well as hear the person making a telephone call. Improvements in technical standards will result in more realistic and reliable sights and sounds.

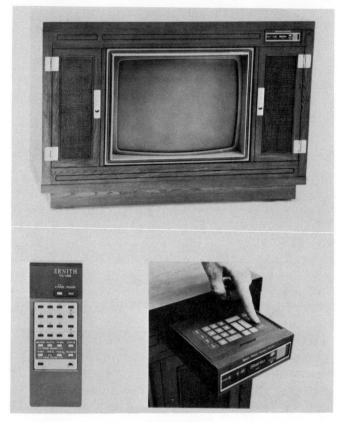

Fig. 26-8. Home entertainment centers are common in many homes. A complete system includes both audio and video components. (Zenith Electronics Corp.)

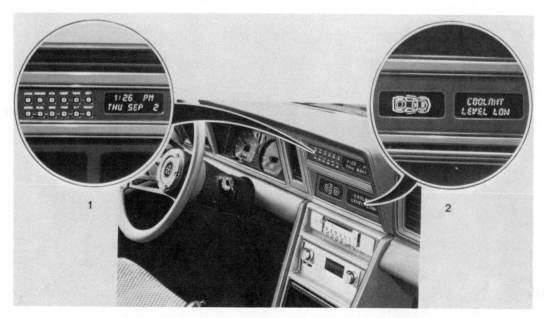

Fig. 26-7. Dashboard displays in cars inform drivers of vehicle function. Displays in this car include a trip timing unit and engine warning lights. (Chrysler Corp.)

PHOTOGRAPHY

Some photography experts doubt that film will even be used in the years to come. Pictures may be recorded on a disc or cartridge. Then, by inserting the container into a television set, the pictures would appear on the screen. Putting the cartridge into a photocopying machine would create prints.

PRINTING

Electrostatic printing has become quite popular over the past 20 years. It is simple and readily available. However, as a printing technique, it is slow. Offset presses can produce thousands more copies per hour. But in the near future, photocopying machines will become more advanced. Four color copiers are being developed that print nearly 40,000 pages an hour.

SATELLITES

Satellites have assisted the television and radio industries for years. Soon, however, other forms of communication will also benefit from these orbiting devices. Telephone calls will be directed around the globe by satellite. Conversations would be routed straight to a satellite and then to another receiver. Underground wires will no longer be necessary. Telephones could then be carried in a pocket or car.

Satellites will also be used in navigation. This is already occurring in marine transportation.

Units are available that track and locate positions anywhere on earth. Travel plans could be checked by locating your car in reference to a satellite. See Fig. 26-9.

COMPUTERS

Computer use will change the nature of home and business life. An example of this is electronic shopping. With this service, a list of products would show on the computer screen. Items would be ordered by typing name, address, and credit card number into a computer linked to the store. Selected merchandise, and the bill, would be sent by computer command. Several retail stores have systems like this in the experimental stage. They only accept telephone/computer orders; no large display area exists. Other stores may begin this service as computer links are completed to more homes.

Another computer service is home education. Computer assisted instruction has already been developed for classroom use. Home education systems are possible by connecting a personal computer to a school's computer system. Students stay home and complete assignments through the computer link, Fig. 26-10.

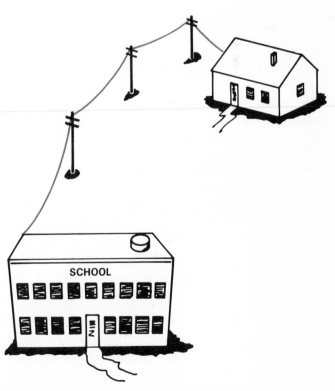

Fig. 26-10. In the future, many students may study at home. A link to the school would allow students to "send" assignments to their teacher. Use of the library would also be possible.

"GROCERY STORE AHEAD TWO BLOCKS, ON THE RIGHT"

Fig. 26-9. Airplanes, trains, and ocean liners are guided by satellites. In the future, cars may also be guided by satellites.

Fig. 26-11. This designer is changing the drawing by touching the screen. Newer computers allow the programmer to "tell" the computer where lines are to be drawn or changed. (General Motors Corp.)

It is thought that computer terminals will eventually take the place of newspapers and magazines. Instead of reading a printed copy of a publication, stories would be displayed on the computer terminal. If you request just the sports, only the sports would arrive at the terminal. Closely related to this procedure are electronic libraries. Books requested on the computer would be displayed on the screen. Many printed documents could be stored in computer data banks for public use.

Worldwide banking systems are nearly possible with current computer and satellite technology. Information about international banking would be relayed to a local bank. This would make foreign trade and tourism easier. The transfer of funds would be simplified by this communication technique.

Finally, computers in the future will be much more friendly, Fig. 26-11. Rather than entering information through a keyboard, the user will talk directly to the computer. The computer will recognize voice commands. Programming languages or typing skills might never have to be learned.

SUMMARY

Technological growth has occurred rapidly in the last few years. This development is known as geometric progression. This skyrocketing progression is known as exponential growth.

Current trends that result from all this growth include electronic billboards, a laser lithographic platemaker, fully automatic cameras, AM stereo, and computer-aided design.

Researchers have quite a lengthy list of what we might expect in the future. Included on this list are televisions with split screens, cameras that do not use film, completely portable telephones, and electronic shopping.

KEY WORDS

All the following words have been used in this chapter. Do you know their meanings?

Computer-aided design, Electronic shopping, Exponential growth, Geometric progression, and Home education.

TEST YOUR KNOWLEDGE — Chapter 26

(Please do not write in this text. Place your answers on a separate sheet.)

1. What is geometric progression?
2. _____ _____ results from inventors building upon the knowledge and improvements of other inventors.
3. List current trends in communication technology of three areas.
4. Only AM signals can achieve a stereo wave. True or False?
5. Researchers make accurate forecasts by reviewing _____ _____ and _____ _____ .
6. Name two future uses for satellites.
7. Explain the concept of electronic shopping.

ACTIVITIES

1. Interview several people over the age of 55. Ask them to describe the changes in communication they have witnessed during their lifetimes. Tape record the interviews.
2. Talk to an office worker at your school or your parents' offices. Ask about new communication devices being used in the office today.
3. Read a science fiction book. Make a list of communication devices mentioned in the book. Are any of these devices actually in use?
4. With your classmates, develop a list of communication devices you would like to see invented. Make illustrations of what these new devices might look like.

GLOSSARY

A

ACRONYM: A word made up of the first letters of a series of words.

ACTIVE SONAR: Sends out and listens for sound waves.

ADVERTISERS: Part of the marketing department in a business, they inform the public of newly available items.

ADVERTISING AGENCY: Organization that specializes in designing promotional material.

AFFILIATES: Local stations of major networks.

AMATEUR BAND: Frequencies used by individuals for long distance personal communication.

AMPLIFICATION: Process of making light, sound, and feelings more intense.

AMPLIFIER: Device that increases the power of a signal.

AMPLIFIER: Part of a stereo system that accepts electrical signals from the turntable, tape deck, or receiver, and powers the signals enough to activate the speakers.

AMPLITUDE: Distance between crest and trough in an electromagnetic wave.

AMPLITUDE MODULATION: Audio signal in which the amplitude of the wave is changed in order to send the message.

ANALOG: Continuous movement.

ANALOG SIGNAL: Signal that varies constantly. It is smooth and continuous.

ANNUAL REPORT: Report concerning the status of a company, including information on stock price ranges, summaries of operation costs, and a note to the stockholders.

ANTENNA: Type of transmission link that aids communication through the atmosphere.

APPLICATION SOFTWARE: Does data processing; solves problems, plays games, gives answers.

APPRENTICESHIP: Time spent observing and working with a skilled worker.

ARCHITECTURAL DRAFTERS: Individuals who specialize in planning homes and buildings. Also known as architects.

ARITHMETIC LOGICAL UNIT (ALU): Part of the CPU, it performs the instructions.

ARTICLES OF DISSOLUTION: Paperwork that must be filed in order to terminate a business.

ARTICLES OF INCORPORATION: An outline of the structure and purpose of a business.

ARTIFICIAL INTELLIGENCE: Machines capable of thinking and making judgements as humans do.

ASA RATING: Indicates the amount of light needed to properly expose film.

ASCII: American Standard Code for Information Interchange. One of two common coding methods used in computer programming.

ASSETS: Equipment, buildings, and property that belong to a business.

ASYNCHRONOUS TRANSMISSION: Serial transmission in which each bit of information is sent by itself.

ATOM: The building block of all substances.

AUDIBLE SOUND: Sound that can be heard by humans.

AUDIBLE SOUND WAVES: Electromagnetic waves in the audio spectrum.

AUDIO SPECTRUM: Electromagnetic waves that are longer than visible light. They cannot be seen, but can be heard.

B

BAKELITE®: Plastic material that is shaped into mold for rubber stamp.

BALANCE: The location of parts within a layout. One of the design principles.

BALANCED ATOM: Atom that has an equal number of protons and electrons.

BASIC: Beginner's All-purpose Symbolic Instruction Code. The most widely used language for personal computers.

BINARY CODE: Method of communication used by computers that uses digital signals.

BLOCK SIGNALING SYSTEM: Light signals that inform railroad engineers of conditions on the tracks ahead of them.

BOARD OF DIRECTORS: Persons who protect the interests of stockholders and make policies for a corporation.

BOUNDED TRANSMISSION CHANNELS: Mediums used in data communication that are affected very little by noise, thereby increasing the distance a signal can travel.

BRAILLE: Printed system of communication designed for the blind. Uses groups of raised dots to represent letters, numbers, etc.

BREAK LINES: Lines that represent a distance of an object that does not change.

BUGS: Problems in a computer program that cause the program to stop working.

BUOYS: Floating device which marks channels and harbors.

BURNING: Process in which the emulsion side of a stencil is exposed to light.

BYLAWS: Document that provides rules about the operation of the company.

C

CABLE: Type of transmission link that allows controlled transmissions.

CALIFORNIA JOB CASE: Case used to store foundry type.

CAMERA: Box that controls the transfer of light to film.

CAMERA READY: Any layout or art that, when photographed as is, will reproduce correctly.

CARTOGRAPHY: Map making.

CARTRIDGE: Part of a stereo tonearm that accepts the vibrations picked up by the stylus and changes them into electrical energy.

CARTRIDGE RECORDER: Type of tape recorder that uses tape with only one loop.

CASSETTE RECORDER: Type of tape recorder that uses cassette tapes.

CATHODE RAY TUBE (CRT): Found in a TV, contains an electron gun that shoots electrons onto the TV screen.

CENTER LINE: Line used to find the center of an object.

CENTRAL PROCESSING UNIT (CPU): Part of the computer where all data is processed and all basic computer functions are done.

CHIEF EXECUTIVE OFFICER (CEO): Directs the company's day-to-day work schedule. Also known as president.

CINEMATOGRAPHY: The science of motion-picture photography.

CITIZEN BAND: Frequencies used by individuals for short distance personal communication.

CLIP ART: Drawn by professional artists, sold in book form. Clipped from book and used in place of original drawing.

COBOL: COmmon Business Oriented Language. Programming language that is used a great deal in business.

CODE: Language computers use to communicate with each other.

CODES: Vehicles for transmitting messages.

COHERENT: Light beams created by waves of a single frequency.

COLOR: Adds emphasis to graphic work. One of the elements of design.

COMMERCIAL BROADCASTS: Broadcasts sent from network stations.

COMMUNICATION: The process of exchanging information.

COMMUNICATION INDUSTRY: Those businesses that engage in the creating and transferring of information.

COMMUNICATION TECHNOLOGY: The process of transmitting information from a source to a destination using codes and storage systems.

COMPANIES: Business enterprises that conduct economic activities.

COMPOSING STICK: Device on which type is placed for composition.

COMPREHENSIVE LAYOUT: Used as a guide by the layout artist. Actual type and art are still not used.

COMPUTER: Electronic device that receives, changes, communicates, and stores information.

COMPUTER-AIDED DRAFTING (CAD): The use of computers to develop drawings.

COMPUTER CHIP: A piece of silicon about 1/4 inch square that contains thousands of transistors.

COMPUTER GRAPHICS: The use of computers to develop drawings.

COMPUTER LANGUAGE: Set of commands that a computer can use to work with data.

CONDUCTORS: Materials that allow easy movement of free electrons.

CONTACT PRINTING FRAME: Developing tool used to make contact sheets.

CONTACT SHEETS: One sheet prints of all negatives from a single roll of film.

CONTINUITY WRITER: Worker who, using information supplied by a business, writes a script.

CONTINUOUS TONE PHOTOGRAPH: Print that contains many shades of gray or variations of colors.

CONTINUOUS WAVE RADAR: Sends out a constant flow of energy. Used to measure the speed of objects.

CONTRAST: Provides a point of emphasis in a layout; can be achieved with colors, text, or lines. One of the design principles.

CONTROL UNIT (CU): Part of the CPU, it directs the processing of each instruction.

CONVENTION: A universal system.

CONVERTER: Device that changes incoming signals to signals that can be used at the destination.

CORPORATION: A business formed by following a specific legal process. Form of ownership in which many people are owners.

CREATION: The assembling or recording of ideas for a message.

CREATIVE LIGHTING: Used to create a certain effect.

CREST: High and low points formed in an electromagnetic wave.

CROPPING: Method of removing unwanted sections of a negative from a print.

CUTAWAY VIEW: Technical illustration designed to show internal details of an object.

CYBERNETICS: Process in which one machine runs another machine.

D

DAGUERREOTYPE: First practical photograph. Its name comes from its inventor, Louis Daguerre.

DAISY WHEEL PRINTER: Type of printer that contains a wheel that rotates into position for printing.

DATA: Unorganized information used in computer programming.

DATA BANKS: Electronic files that can be retrieved using a computer.

DATA COMMUNICATIONS: The process of transmitting information in binary form between two points; a system that allows computers to talk to one another.

DATA PROCESSING: Process of entering information into a computer, giving instructions to follow, and receiving results.

DATA PROCESSORS: People who enter information into a computer, give the computer instructions to follow, and obtain results.

DEBUGGING: The process of correcting bugs in a computer program.

DESIGN AGENCIES: Companies that specialize in the creative construction of commercial items. Also known as design firms.

DESIGN FIRMS: Companies that specialize in the creative construction of commercial items. Also known as design agencies.

DESIGNING: The first step toward a completed graphic design.

DESIGNING PROCESS: Steps followed to create a message. Process includes ideation, purpose, and creation.

DESIGN PHASE: Portion of the programming process in which the programmer creates the best program method and then puts it to use.

DESIGN PRINCIPLES: Describe the nature of the layout. They are: balance, contrast, rhythm, proportion, and unity.

DETAILED FLOW CHART: Type of flow chart used to give specific points of a program.

DETECTION: The location of an object within the path of a beam.

DIAPHRAGM: A flexible elastic membrane found in a telephone. It vibrates in the exact way the original sound is vibrating.

DIGITAL SIGNALS: Signals that can only represent one of two states: on or off.

DIMENSION LINE: Line used to show the dimensions of an object.

DIRECT BROADCAST SATELLITE: Satellite that broadcasts directly to a home.

DIRECT FEEDBACK: Spoken or written words that show a message has been received.

DIRECT-IMAGE PLATE: Type of lithographic plate in which the master stencil is made with special pencil or pen on a paper plate.

DISTRIBUTED NETWORKS: Network in which a central computer is attached to other computers.

DISTRIBUTION UNIT: Part of the marketing department of a business, they deliver items to customers.

DITTO PRINTING: Printing procedure in which words or art are done on a special master.

DITTO PRINTING PROCESS: Printing process in which the original design is typed or drawn directly on a master sheet.

DOPPLER EFFECT: The change in frequency of a sound source as it approaches and then passes another object.

DOT MATRIX PRINTER: Printer that forms letters and symbols with tiny dots of ink.

DOWNLINK: Signal that travels from a satellite back to the earth's surface.

DRAFT: An outline of a message.

DRAFTING: Communication method that uses drawings made with instruments to present ideas.

DRAFTING BOARD: Work surface for drafting.

DRAWINGS: The fastest, easiest, and most common method of graphic communication.

DRY POINT ETCHING: Form of intaglio printing in which lines are scratched on clear plastic sheets. When surface is covered with ink, scratched lines fill with ink.

E

EBCDIC: Extended Binary-Coded-Decimal Interchange Code. One of two common coding methods used in computer programming.

ELASTICITY: The ability of a mass to return to its original shape after an outside force has been removed.

ELECTRIC CURRENT: The flow of electrons.

ELECTRICITY: The movement of electrons from one atom to another. Occurs in order to create balanced atoms.

ELECTROMAGNETIC FIELD: Area of magnetic energy produced by electric current.

ELECTROMAGNETIC SPECTRUM: Range of all possible electromagnetic waves.

ELECTROMAGNETIC WAVES: Created by an oscillating electromagnetic field. These waves transmit electronic messages.

ELECTROMECHANICAL SWITCHING: Telephone switching done using electronic impulses.

ELECTRONIC COMMUNICATION: The use of electrical energy to transmit information between individuals or systems.

ELECTRONIC SWITCHING SYSTEM: Device that routes calls from source to destination using computers.

ELECTRONS: Negatively-charged particles contained within the nucleus.

ELECTROSTATIC COPYING: Printing procedure in which powdered ink is fused to paper. Also kown as photocopying.

ELEMENTS OF DESIGN: Those special considerations necessary to create a graphic message. They are: shape, mass, lines, texture, and color.

ELLIPSE: An oval shape that results when drawing a circle in an isometric view.

ENGINEERING DRAWINGS: Designs of consumer products.

ENLARGEMENTS: Photographs developed from negatives.

ENTERPRISE: Business organization or company.

EXECUTABLE INSTRUCTIONS: Program instructions that tell BASIC what to do while a program is running.

EXPLODED VIEW: Technical illustration that describes the manner in which an object is put together.

EXPONENTIAL GROWTH: Steady growth pattern.

EXTRATERRESTRIAL: Messages received from outside the earth.

EYE LEVEL: View positioned horizontally, in front of the eyes.

F

FEDERAL COMMUNICATIONS COMMISSION (FCC): Group that regulates radio, TV, and telephone communication.

FEDERAL COMMUNICATIONS COMMISSION (FCC) CATEGORIES: Division of telecommunication systems, designed by the FCC, for users of various systems.

FEEDBACK: A sign that a message has been received.

FIBER OPTICS: Transmission of messages as light signals along a transparent (clear) fiber.

FIXED SONAR: Transducer attached to a vessel. Sound waves are sent out and received in the same location.

FLAT: A ruled goldenrod sheet.

FLOW CHART: Common programming technique that involves mapping out the logic needed in a program.

FLUCTUATE: Change in energy level.

FORESHORTENING: Principle that states the farther away an object is viewed, the smaller it will appear.

FORMAL BALANCE: Achieved when parts are centered in a design. Each item is orderly and evenly weighted.

FORMAT: Types of programs a station will provide to the listening audience.

FORTRAN: FORmula TRANslator. Programming language used widely for science and engineering programs.

FOUNDRY TYPE: Form of relief printing invented by Johann Gutenberg.

FREELANCE PHOTOGRAPHERS: Photographers who work for themselves.

FRENCH CURVE: Drafting tool used to draw curved lines.

FREQUENCY: The number of electromagnetic waves that pass a given distance in one second.

FREQUENCY MODULATION: Audio signal in

which the frequency of the wave is changed in order to send the message.

F-STOP: On a camera, the shutter opening that determines the amount of light that strikes the film.

G

GAMMA RAYS: The shortest waves in the electromagnetic spectrum.

GENERAL FLOW CHART: Type of flow chart that gives major functions and logic needed in a program.

GENERAL LEDGER: Official record of all financial matters.

GENERAL MANAGER: Person who is the leader of all personnel. Also known as a station manager.

GENERATIONS: Divisions of time for computer development.

GEOMETRIC PROGRESSION: Growth that occurs rapidly instead of at a steady pace.

GEOSYNCHRONOUS ORBIT: Position at which a satellite will orbit the earth at the same speed the earth revolves on its axis.

GOLDENROD SHEET: Special yellow paper used when preparing lithographic plate. It is heavy and opaque, blocking out light in non-printing areas.

GRAPHIC COMMUNICATION: The use of printed images to communicate.

GRAPH PAPER: Lined paper used to draw straight, parallel, or equally spaced lines.

GRAVURE PRINTING: Printing procedure in which ink is transferred from an engraved image to another surface. Also known as intaglio printing.

GRAY SCALE: Strip of film with a range of grays from clear to black. Helps in exposing film for the proper time.

GRID SYSTEM: Process used to change the size of a sketch.

GUIDELINES: Line used to give a sketch its basic shape.

GUIDES: Mark the base in order to position each sheet in the same place.

H

HALFTONE NEGATIVE: A photograph that is converted into a series of dots.

HARD COPY: The output of a printer or plotter.

HARDWARE: Equipment that makes up an entire computer system.

HARD-WIRED SYSTEM: System or equipment permanently connected by wire.

HEAT TRANSFER PRINTING: Printing procedure in which images are lifted from a carrier sheet to a transfer surface using heat.

HIDDEN LINE: Line used to show edges or parts of an object that are not visible.

HIEROGLYPHICS: Use of pictures to communicate ideas.

HIGH-RESOLUTION: Computer screen or printer having a large number of dots per inch, allowing for clear, detailed drawings and letters.

HOLOGRAPHS: Three-dimensional images created with light.

HOUSEKEEPING: In utility software, a job consisting of clearing storage areas, starting programs, and storing data for later use.

HUMAN-TO-HUMAN COMMUNICATION: Exchange of information which only involves people. It is the simplest form of exchanging information.

HUMAN-TO-MACHINE COMMUNICATION: Exchange of information from people to devices (typewriters, telephones, power tools).

I

IDEATION: Getting an idea.

ILLUMINATION: Lighting placed on streets, billboards, storefronts, etc. during night hours.

INAUDIBLE SOUNDS: Sounds that cannot be heard by humans.

INCLINED LINE: Lines drawn at an angle that are not completely parallel with the edge of a sheet of paper.

INDEPENDENT STATIONS: Stations not owned by a major network.

INDIRECT FEEDBACK: The observation of later actions that show a message has been received.

INDUSTRY: A group of related companies.

INFORMAL BALANCE: Achieved when the arrangement of parts is random.

INFORMATION OVERLOAD: Exposure to too much information.

INK JET PRINTER: Printer in which images are formed by droplets of ink arranged to form letters and symbols.

INK JET PRINTING: Printing procedure in which drops of ink are shot onto paper.

INPUT: Resources that go into making a product.

INPUT DEVICES: Any piece of hardware used to input data into a computer.

INPUT/OUTPUT DEVICES (I/O DEVICES): Peripherals that allow data to be put into or taken out of the computer.

INSULATORS: Materials that do not allow easy movement of free electrons.

INTAGLIO PRINTING: Printing procedure in which ink is transferred from an engraved image to another surface. Also known as gravure printing.

INTEGRATED CIRCUITS: Complete circuit systems manufactured on a single silicon chip.

INTERFACE: A group of rules that control the way in which two machines or processes interact.

INTERFERENCE: A distortion of signals being transmitted to a receiver.

IONOSPHERE: Layer of the earth's atmosphere, which reflects radio signals back to earth.

ISOMETRIC TEMPLATE: Tool used to draw an ellipse.

ISOMETRIC VIEW: Pictorial view drawn as it appears to the eye. Lines are shown full length, or in proportion.

K

KINETOSCOPE: A machine used in the past to view moving pictures.

KROY MACHINES: Printing device that stamps letters and symbols onto clear, gummed tape.

L

LANGUAGE TRANSLATORS: Programs that change languages inside the computer.

LASER: A form of radiation that is amplified to a high energy level. LASER stands for Light Amplification by Stimulated Emission of Radiation.

LATENT IMAGE: Invisible image.

LAYOUT: The assembly of copy and artwork.

LEADING: Used to insert space between lines of foundry type.

LIGHT EMITTING DIODE (LED): Light source that is visible.

LINE OF SIGHT TRANSMISSION: A message that, when sent through the atmosphere, is able to travel in a straight line only.

LINES: Strokes made with pens, pencils, or tape. One of the elements of design.

LITHOGRAPHY: Procedure that permits the printing of images from flat surfaces. Also known as offset printing.

LOUDNESS: Makes up the power of a sound; is determined by the pressure (strength) of a wave.

LOW-RESOLUTION: Computer screen or printer having a small number of dots per inch, allowing for less detailed output than a high-resolution system.

M

MACHINE-TO-HUMAN COMMUNICATION: Exchange of information from device (computer, stereo, etc.) to people.

MACHINE-TO-MACHINE COMMUNICATION: Exchange of information between machines.

MAINFRAME COMPUTER: Computer with a large memory. Used mostly for general purpose work.

MANAGERS: Personnel who set goals for the enterprise, direct the daily functions of the company, and rate work in progress.

MASKING OFF: Blocking an area so that unwanted printing is avoided.

MASS: The amount of space used in a graphic design. One of the elements of design.

MASS: The quantity of matter that forms a substance.

MASTER PLATE: Original design used for the printing process.

MASTER PRINTERS: Experienced workers in the printing trade.

MATRIX: Material used to make a rubber stamp.

MECHANICAL LAYOUT: Completed design that will be used as the master for future production work.

MECHANICAL LEAD HOLDER: Tool used in drafting that rarely needs sharpening. Used when sharp, clean lines are required.

MEDIUMS: The physical equipment used to send a message.

MENU PAD: Piece of equipment used to enter commands into a computer.

MICROCOMPUTER: Small, self-contained computer. Also known as a personal computer.

MICROFICHE: Copy of data reduced onto a four inch by six inch filmsheet.

MICROFILM: Copy of data reduced onto 16mm continuous film.

MICROPHONE: Piece of electronic equipment used inside TV and radio studios to convert voices and music into electrical signals.

MICROPROCESSORS: Small silicon wafers that replaced wires and tubes of earlier computers.

MICROWAVE LANDING SYSTEM (MLS): Acoustical communication device that uses microwaves to aid aircraft while landing.

MICROWAVES: Wavelengths of short frequency.

MIMEOGRAPH PRINTING: Printing procedure in which ink is forced through the openings on a master stencil.

MIMEOGRAPH PRINTING PROCESS: Printing process in which ink is forced through a master stencil onto paper.

MODULATION: The process of controlling a carrier wave.

MOMENTUM: The building of strength and speed.

MULTIVIEW DRAWING: An illustration that requires more than one view to properly describe the shape of the object being drawn.

N

NEEDLE: Hard material, usually diamond, located at the end of the stylus.

NEGATIVE: Image produced after developing film.

NEON LIGHTS: Glass tubing shaped as letters and designs. Colored light is produced by using neon gas.

NETWORK: Worldwide system that provides uniform service to the entire world.

NETWORK: Data communications systems organized into a group.

NEUTRONS: Particles that have neither a positive nor negative charge and are contained within the nucleus.

NOISE: Interference that distorts a message. Any signal not present in the original message.

NON-EXECUTABLE INSTRUCTIONS: Instructions that cause no change in program flow or in the running of the program.

NONIMPACT PRINTER: Type of printer in which no part of the printer touches paper.

NUCLEUS: Center of an atom.

O

OBJECTIVE VIEWPOINT: Camera angle in which the camera watches the action.

OBJECT LINE: Line used to show the visible edges of an object.

OBLIQUE VIEW: Pictorial view containing a front view that is parallel to the picture plane and shown as true shape and size.

OFFSET PRINTING: Procedure that permits the printing of images from flat surfaces. Also known as lithography.

ONE-WAY RADIOS: Radios that only send messages or receive messages between a source and destination.

OPERATING PROGRAM: Software that enables the user to make lines, circles, and labels on the screen.

OPERATING SYSTEM: Software used to control the computer and its input/output and storage devices.

OPTIC WAVE GUIDE: Cable made up of glass fiber.

ORTHOGRAPHIC DRAWING: The system of organizing views to describe a three-dimensional object in two dimensions.

ORTHOGRAPHIC PROJECTION: Procedure used to develop several views of a single object. The projection image is viewed from an infinite distance.

ORTHOGRAPHIC VIEW: View seen from the point of reference in an orthographic projection.

OSCILLATE: Rapidly change directions back and forth.

OSCILLATOR: Device that produces repeating signals.

OUTPUT: Final product or service resulting from an enterprise's work.

OUTPUT DEVICES: Any piece of hardware that allows the user to get information out of a computer.

OVERLAPPING: A condition in which objects viewed from a distance appear blocked or to be blocking other objects in the view.

P

PARALLEL TRANSMISSION: Single pieces of information sent between computer and terminal on their own wire, at the same time.

PARTNERSHIP: Business owned by two or more people.

PASSIVE SONAR: Used to listen for sounds.

PASTEUP: Process in which copy is set in type and artwork is put in place.

PERIPHERALS: Attachments for computer hardware.

PERPENDICULAR VIEWING PLANE: Viewing plane that shows proper height and width, but does not show depth.

PERSONAL COMPUTER: Small, self-contained computer. Also known as a microcomputer.

PERSONNEL: Workers at a business.

PERSPECTIVE DRAWING: Pictorial drawing that shows an object as it appears to the eye from a certain location. Provides a realistic view of a large object.

PHONOGRAPH: Communication device that records and plays back messages. Invented by Thomas A. Edison.

PHOTOCOPYING: Printing procedure in which powdered ink is fused to paper. Also known as electrostatic copying.

PHOTO-DIRECT PLATE: Process in which the master stencil is made on a light-sensitive paper plate.

PHOTOGRAPHIC COMPOSITION: Arranging or putting pictures together in the viewfinder before shooting the picture.

PHOTOTYPESETTING: Type made photographically on light-sensitive film.

PICTORIAL SKETCH: Drawing that shows the height, width, and depth of an object. This type of sketch is three-dimensional. Also known as a pictorial drawing.

PITCH: Makes up the power of sound; is determined by frequency of a wave.

PL/1: Programmer's Language 1. Used for science and business programs.

PLATEN: Smooth, flat plate used for printing.

PLAYBACK HEAD: Device located in a recorder used to retrieve sound recordings.

PLOTTER: Computer hardware that creates hard copy of information.

POINT OF REFERENCE: Established point from which an orthographic projection is made.

PRESENSITIZED PLATE: Process in which master stencil is made on aluminum plate. Plate is treated on both sides, allowing image to be developed on either side.

PRESIDENT: Directs the company's day-to-day work schedule. Also known as chief executive officer (CEO).

PRESS MAKE-READY: Preparation of printing press systems that will be used during printing.

PRINTERS: Computer hardware that creates hard copy of information.

PRINTING: A two-step process followed to make a photographic print.

PROCESSING: Developing a roll of film into a series of negatives.

PRODUCTION: The process of getting all needed items together in order to put a show on the air.

PROGRAM CLOCK: Daily schedule of programs, news, commercials, and sports shown as a clock divided into 60 minutes.

PROGRAM DESIGN: The path a program follows to reach the desired outcome.

PROGRAM DESIGN: The path a program follows to reach the desired outcome.

PROGRAMMERS: People who write hardware instructions or software for computers.

PROGRAMMING: Writing code for a computer.

PROGRAMMING: The planning and producing of shows.

PROGRAMMING TECHNIQUES: Tools used to help computer programmers design software.

PROGRAMS: A set of instructions used to run a computer. Also known as software.

PROJECTED IMAGERY: Using focused light to project images on a screen.

PROPAGANDA: The use of false or misleading information to form a certain opinion.

PROPORTION: The relationship of physical sizes between objects in a design. One of the design principles.

PROPRIETOR: Owner of a proprietorship.

PROPRIETORSHIP: Business owned by just one person.

PROPS: Items actually used during the acting portion of a show.

PROTON: Positively-charged particle contained within the nucleus.

PUBLISHING HOUSES: Firms that print newspapers, magazines, and books.

PULSE RADAR: Sends out timed bursts of energy. Distance and direction of an object are measured by these pulses.

Q

QUADS: Used to space within a line of foundry type.

QUICK PRINTERS: Printing firm that provides service while you wait.

R

RADAR: Form of acoustical communication that uses inaudible sound waves to detect solid objects in the path of a wave.

RANDOM ACCESS MEMORY (RAM): Contains information for use by the CPU and stores this information while the computer is turned on.

READ-ONLY MEMORY (ROM): Memory which can only be read by the CPU, it cannot be erased or changed.

RECEIVER: Part of a stereo system that tunes in radio signals. Also known as a tuner.

RECORDING HEAD: Coil that records electrical signals onto a tape.

RECORDS: Vinyl discs that contain a series of grooves. The grooves represent waves of pressure.

REFINED SKETCH: A drawing that has more lines and details than a thumbnail sketch.

REFLECTION: Earliest method of light communication. Shiny objects were used to reflect the sun's rays.

REGISTRATION MARKS: Used to position a stencil consistently with the transfer medium.

RELIEF PRINTING: The transfer of images from a raised surface.

RENDERING: A simple isometric view with shaded areas.

REPORTORIAL VIEWPOINT: Camera angle in which the speaker looks directly into the camera.

RESERVED WORDS: Words that have special meaning in certain computer languages.

RESISTANCE: Opposition to electrical current.

RESISTORS: Devices that create resistance.

RESOLUTION: Number of dots that can be drawn per inch.

RHYTHM: The repetition of certain aspects of a design. Serves to guide the eye along a design. One of the design principles.

ROUGH LAYOUT: A layout that is more accurate and detailed than a thumbnail sketch. Used to check the appearance of the final design.

S

SALES COMMISSION: A monetary share of a sale.

SALES FORCE: Part of the marketing department of a business, they complete the task of selling goods or services to potential buyers.

SCHEMATIC DRAWINGS: Technical drawings that show the inner circuitry of electronic devices.

SCREEN PROCESS PRINTING: Printing procedure in which ink is forced through a prepared screen. Also known as silk screening.

SCRIPT WRITERS: Personnel who prepare a script or copy for broadcast.

SERIAL TRANSMISSION: Bits of information sent between computer and terminal one at a time, over one wire.

SET DESIGN: The organization of all equipment, scenery, and props needed for a show.

SET DRESSINGS: Any items that might be found in the location being created on a stage.

SETTINGS: The large pieces of scenery that make up the areas of the set.

SHADE: The dark side of an object.

SHADOW: Darkened area caused by blocked light.

SHAPE: A combination of lines and mass. One of the elements of design.

SIGNING: The use of hands and fingers to communicate with words. Used by the deaf and for the deaf.

SILICON CHIP: Small printed circuit boards made on silicon or germanium crystals.

SILK SCREENING: Printing procedure in which ink is forced through a prepared screen. Also known as screen process printing.

SLUGS: Used to insert space between lines of foundry type.

SOFTWARE: A set of instructions used to run a computer. Also known as programs.

SONAR: A form of acoustical communication that uses reflected sound waves to send signals through water, to detect underwater objects, or to determine ocean depths.

SONUBUOY: A transducer with a collar that keeps it afloat in water.

SOURCE: The starting point of messages to be sent.

SPACERS: Used to make space within a line of foundry type.

SPEAKERS: Part of a stereo system that changes electrical energy back into acoustical energy.

SPECIAL PURPOSE PRINTERS: Printing firms that produce legal documents and related materials.

SPOT LIGHTING: Used to brighten a set.

SPREADSHEET: A grid made up of rows and columns.

SQUEEGEE: A tool used to evenly spread and then force ink through a stencil.

STAFF: People hired to do specific jobs.

STAGING: Putting together of all items needed to create the feeling of a certain place.

STATIONARY HEAD: Nonmoving portion of a tape deck that records or plays tapes.

STATION MANAGER: Person who is the leader of all personnel. Also known as a general manager.

STENCIL: A master design through which ink or paint is forced.

STEREOPHONIC RECORDINGS: Recordings made using several microphones.

STOCKHOLDERS: Owners of corporations. They own shares of the business.

STORAGE: Keeping a message for later use.

STORYBOARD: An outline for audio-visual presentation. A series of rough sketches with video and audio directions given underneath each sketch.

STUDIO: Area in which TV and radio broadcasts are created; the center of activity and attention.

STYLUS: Part of a stereo tonearm that picks up signals from the groove of a record.

SUBJECTIVE VIEWPOINT: Camera angle in which the camera takes the view of the actor or actress.

SUBSCRIBER: Household that rents cable service.

SUPPLEMENTAL COMMUNICATION SYSTEMS: Exchanges of information that do not fit into the four general communications groups.

SURVEYORS: People who take measurements of terrain to be used in cartography.

SWITCHES: Direct the flow of electricity.

SWITCHING SYSTEMS: Route calls from source to destination.

SYMMETRY: Being exactly the same on both sides or ends.

SYNCHRONOUS TRANSMISSION: Serial transmission in which bits of information are sent in groups.

SYSTEM SOFTWARE: Contained within the computer system. Controls CPU functions. Translates languages.

T

TAKE-UP REEL: Part of an open-reel recorder that collects audio tape.

TAPES: Plastic strips coated with a magnetic material. Used to record sound waves.

TARGETING: Directing messages to a particular

group of people.

TECHNICAL COMMUNICATION SYSTEMS: Technologies that use technical devices or systems to transmit messages.

TECHNICAL GRAPHICS: Visual images created through the use of tools and instruments.

TECHNICAL ILLUSTRATION: Pictorial drawing that is shaded and detailed to give a realistic view of the object. Describes a technical device or system.

TELECOMMUNICATIONS: Transmitting information between distant points.

TELEGRAPH: Communication device that uses electrical current to spell out words, which are then set over wires.

TELEPHONY: Signals that are not coded but are easily recognized by the listener.

TELETEXT: System that transmits words and graphics through regular television signals.

TEMPLATE: Drafting tool used to draw circles and make curved lines.

TEST SHEET: Print paper that is exposed in sections for various time intervals. It is used to determine the proper exposure time.

TEXTURE: The amount of roughness or smoothness of an object. One of the elements of design.

THUMBNAIL SKETCH: Basic sketch that shows simple shapes with only a few lines.

TIME-SHARING NETWORK: Network in which any number of terminals are connected to one large central computer.

TOLERANCE: The total amount a design part is permitted to vary from the stated size.

TONEARM: Part of a stereo turntable that changes signals from the grooves of a record into electrical energy.

TRACK: A cut in the wall of a groove on a record. This cut is the sound wave from the microphone.

TRANSDUCER: A device that acts like an underwater microphone and loudspeaker. Picks up and sends out sound waves.

TRANSISTOR: Device that either controls or amplifies electronic signals. Powerful amplifier.

TRANSMITTER: Device used to start a message on its way.

TRANSMITTING: Sending of a message.

TROUGH: Lowest point of an electromagnetic wave.

T-SQUARE: Common drafting tool used to draw straight, parallel lines.

TUNER: Part of a stereo system that receives radio signals. Also known as a receiver.

TWEETER: Part of a speaker that delivers high-frequency sounds.

TWO-DIMENSIONAL SKETCH: Drawing that shows only the height and width of an object.

TWO-WAY RADIOS: Radios that both send and receive messages. Feedback is possible.

U

UNBOUNDED TRANSMISSION CHANNELS: Mediums used in data communications that are affected by noise and need repeaters in order to travel a distance.

UNITY: Pulls a design together. One of the design principles.

UPLINK: Signal that travels from a ground station on the earth's surface to a satellite.

UTILITY SOFTWARE: Cleans house, measures efficiency of system, and does routine programs.

V

VANISHING POINTS: Those spots where lines of sight are developed.

VICE-PRESIDENT: Person who watches over the work done in a particular group.

VIDEOTEXT: System that transmits information through telephone lines to be displayed on a screen.

VIEWFINDER: Part of a camera used to select scenes.

VIEWING PLANE: Angle at which an object is seen. Used when developing a drawing.

VISUALIZATION: The process of thinking through different solutions to solve a problem or develop a design.

VOLTAGE: Force supplied to produced electricity.

W

WAVELENGTH: Distance between the crest of one electromagnetic wave and the crest of the next wave.

WELDING DRAWINGS: Serve as instructional plans for placement of welds.

WIRELESS: Term originally used for radio; a device that transmits sound without the use of wires.

WHITEPRINTS: Engineering drawings having a white background and blue lines.

WOOFER: Part of the speaker that delivers low-frequency sounds.

WORD PROCESSORS: Computerized, electronic typewriters.

INDEX